CONVERGENCE AND UNIFORMITY IN TOPOLOGY

BY

JOHN W. TUKEY

PRINCETON
PRINCETON UNIVERSITY PRESS
LONDON: HUMPHREY MILFORD
OXFORD UNIVERSITY PRESS
1940

PRINTED IN U.S.A.
Lithoprinted by Edwards Brothers, Inc., Lithoprinters
Ann Arbor, Michigan, 1940

ANNALS OF MATHEMATICS STUDIES

NUMBER 2

INTRODUCTION

This is an attempt to give a convenient, systematic, and natural treatment of some of the fundamentals of topology--the fundamentals of a topology which will be a tool of general application to other branches of mathematics as well as an interesting and satisfactory theory.

In this treatment we show that, theoretically and practically, convergence is a notion of central importance in topology.

The results set forth here are not all new; there are old results with old proofs, new results with old proofs, old results with new proofs, and new results with new proofs.

In large part this material is drawn from my doctor's thesis (Princeton University, 1939). I take this oppostunity to thank the many friends who have discussed various related questions with me. In particular, A. H. Stone, B. McMillan and B. Tuckerman have read the manuscript and made many helpful suggestions.

CONTENTS

		Page
General Usage of the Alphabets		vii
Special Usage of the Alphabets		viii
Usage of General Symbols		ix
Chapter I	Ordering	1
Chapter II	Direction	10
Chapter III	Convergence	16
Chapter IV	Compactness	31
Chapter V	Normality	43
Chapter VI	Structs	55
Chapter VII	Function-Spaces	71
Chapter VIII	Examples	78
Chapter IX	Discussion	82
Bibliography		87
Indices		89

GENERAL USAGE OF THE ALPHABETS

Small Latin Letters: a, b, c, $\cdots$, are used for individuals (indices, elements, points, etc.).

Latin capitals: A, B, C, $\cdots$, are used for sets, collections, etc.

German capitals: $\mathfrak{B}$, $\mathfrak{H}$, $\mathfrak{L}$, $\mathfrak{M}$, $\mathfrak{N}$, $\mathfrak{U}$, $\mathfrak{V}$, $\mathfrak{W}$, are used for collections of sets, in particular for coverings.

Script capitals: $\mathcal{A}$, $\mathcal{B}$, $\mathcal{C}$, $\cdots$, are used for ordered systems. Except in Chapter I, these ordered systems are assumed to be directed.

Small Greek letters: α, β, γ, δ, ε, $\cdots$ ν, are used for finite sets. φ and ψ are used for functions.

Greek capitals: Γ, Δ, $\cdots$, are used for collections of finite sets, when these finite sets are denoted by small Greek letters.

Dashed small Greek letters: $\bar{\alpha}$, $\bar{\beta}$, $\bar{\gamma}$, $\bar{\delta}$, $\cdots$, are used for sets which are regarded as elements and which need not be finite.

Dashed Greek capitals: $\bar{\Gamma}$, $\bar{\Delta}$, $\cdots$, are used for collections of sets denoted by the corresponding dashed small letter.

(An exception to the last four rules occurs in Chapter VIII.)

SPECIAL USAGE OF THE ALPHABETS

Corresponding letters in different alphabets are used for related objects.

D is frequently used to denote 2^X considered as a set rather than as a set of sets.

The indices i, j, k, l, m, and n usually run thru the positive integers.

N is used frequently in two distinct senses; as the set of all positive integers, or as a nbd of the point being considered.

U , V , and W are open sets.
$\mathfrak{U}$, $\mathfrak{V}$, and $\mathfrak{W}$ are open coverings.

These conventions are usually independent of indices.

Examples.

a_1 is usually an element of A . γ' is usually a finite subset of C . δ belongs to Δ (or perhaps to $\Delta_1 \subset \Delta$).

USAGE OF GENERAL SYMBOLS

$\in$ denotes class-membership.

$\subset$ and $\supset$ denote set-theoretical inclusion.

$\cap$ and $\cup$ denote set-theoretical intersection and union.

$<$, $>$, $\wedge$ and $\vee$ denote the corresponding lattice operations.

$\{ \}$ are used to form sets. Thus, $\{U\}$ is the family of all U's. Also $\{x'\}$ is often the set made up of x' alone.

$|$ is used to indicate domains, etc. It can usually be read "where" or "running thru". Thus $f(x|X)$ is a function defined for x in X , and $\{f(x)|x \in X\}$, often written $\{f(x)|X\}$ is the set of its functional values.

$\emptyset$ denotes the empty set.

Propositions whose proof is left to the reader are designated by "****" rather than by the words "Lemma" or "Theorem".

Chapter I

ORDERING

1. Introduction
2. Set Theory
3. Orderings
4. Functions
5. Products
6. Zorn's Lemma
7. Cardinal Numbers

1. Introduction. In this chapter we concern ourselves chiefly with definitions and notation for sets, functions, and ordered systems. We assume familiarity only with the most elementary facts about sets and functions.

Definitions and notation must represent a compromise between theory and practice. We tend to use weaker definitions and more precise notation than some authors. In places we follow the definitions and notation used by N. Bourbaki and his collaborators.

Certain conventions of notation, which allow the notation to carry implicit information, and which are used consistently thruout, are laid down in this chapter.

We begin with our definitions and notation for set theory (§2), ordered systems (§3), and functions (§4). These are applied to products of ordered systems (§5) and Zorn's Lemma (§6). We conclude with a modicum of the theory of cardinal numbers (§7).

2. Set Theory. We assume that the reader is familiar with the elements of set theory, and concern ourselves with notation.

The set of elements x which have property P is $x|P$. This is often further abbreviated in obvious ways, as $\mathfrak{U}_a|A$ for $\mathfrak{U}_a|a \in A$.

This last abbreviation is an example of an important convention. We use small letters, a, b, c, $\cdots$ for points belonging to sets denoted by the corresponding capital letters, A, B, C, $\cdots$. For example, if B is a subset of a set A of real numbers, then "For every a and some b, $b > a$" means "For every element a of A and for some element b of B (which may depend on a, otherwise we would say "for some b and every a") $b > a$".

We often use the letters i , j , k , l , m , n , for integers and N for the set of all positive integers.

We consider a set, its points, and its subsets. If the point a belongs to the set A we write $a \in A$ or $A \ni a$. (We have no hesitation in inverting any relation symbol.) If every point of A belongs to B we write $A \subset B$ or $B \supset A$ and say that A is a subset of B . The empty set is $\emptyset$ (pronounced as the phonetic symbol and Scandinavian vowel). We make no formal distinction between the empty subsets of different sets. The points belonging to either A or B or both form the <u>union</u> $A \cup B$ of A and B . (We have adopted $\cap$, $\cup$, "intersection" and "union" rather than the older $+$, $\cdot$, Σ , Π , "product" , and "sum" to avoid confusion when the sets belong to an algebraic entity (group, linear space, etc.). The points common to A and B form the <u>intersection</u> $A \cap B$ of A and B .

The union of a collection of sets $A_h | H$ is denoted variously as $\cup_h A_h$, $\cup\{A_h | H\}$, and $\cup_h A_h | H$. The intersection of $A_h | H$ is treated similarly. Two particular cases are $\cup\{A_h | \emptyset\} = \emptyset$ and $\cap\{A_h | \emptyset\} = X$, where X is the set, with whose subsets we are operating.

The points in A but not in B form the set $A-B$ (This notation is not very satisfactory). The set of all subsets of C is 2^C .

If $A \cap B \neq \emptyset$ we say that A <u>meets</u> B ; if $A \cap B = \emptyset$ we say that A and B are <u>disjoint</u>. A collection of sets is disjoint if each pair of them is disjoint.

We will often use alternative terms for "point" , "set" , and "subset" , in an effort to avoid confusion. Elements, indices, objects, etc., may belong to or be contained in families, classes, systems, aggregates, or collections. The prefix <u>sub</u> is used in the same sense when applied to all these terms. A typical usage is, "The system of all families of subsets of A ."

If, on the other hand, we apply the prefix <u>sub</u> to a word meaning a set with an attached "structure" of some sort, e.g., "space" , "ordered system" , "group" , we imply that the set concerned is a subset with the "structure" induced by the "structure" of the object of which it is a subset.

<u>3. Orderings</u>. We consider a set A and a binary relation $>$ defined on the set (more precisely, perhaps, defined on the product of the set by itself). That is, for each ordered pair a' , a'' of elements of A , one of the incompatible formulas $a' > a''$ or $a' \not> a''$ is accepted. (We denote negation by cancellation with /). If the relation is <u>transitive</u>, that is, if we have

3.1 $a' > a''$ **and** $a'' > a^*$ **imply** $a' > a^*$,

then we say that $\mathfrak{a} = (A,>)$ is an ordered system. (We use "element" and "system" when discussing sets with an ordering relation.)

An ordered system may have several important properties. It is linear if either $a' > a''$ or $a' < a''$ or both, for every pair of distinct elements a' , a'' . It is properly ordered if $a' > a''$ and $a' < a''$ imply $a' = a''$ (that is, a' and a'' are the same). It is reflexive if $a > a$ for all a . It is irreflexive if $a \not> a$ for all a . It is symmetric if $a' > a''$ implies $a'' > a'$. It is trivially ordered if $a' > a''$ for all a' and a'' . It is vacuously ordered if $a' \not> a''$ for all a' and a'' .

If B is a subset of A , and $>$ is defined on A , then it is defined on B ; if $>$ is transitive on A , then it is transitive on B . Hence, if $\mathfrak{a} = (A,>)$ is an ordered system, then $\mathfrak{B} = (B,>)$ is an ordered system. We call $\mathfrak{B}$ a subsystem of $\mathfrak{a}$, implying thereby that the ordering relation of $\mathfrak{B}$ can be obtained from that of $\mathfrak{a}$ in this way.

Two ordered systems are isomorphic if there is a one-to-one correspondance between them which carries their order relations into each other. We write $\mathfrak{a} \cong \mathfrak{B}$ for this relation. It is clear that all the notions discussed in this § are invariant under passage to an isomorphic system.

Associated with each set there are two ordered systems: the vacuously ordered system made up of the points of the set, and the system of all subsets of the set ordered by $\supset$. There is a logical distinction between a set and a vacuously ordered system, but we shall have no occasion to insist upon it.

The second ordered system and its subsystems are of extreme importance. If we speak of a system of subsets of some set, we will assume, unless we specifically indicate the contrary, that the system is ordered by $\supset$.

We denote a subset of C by $\bar{\gamma}$ and a collection of $\bar{\gamma}$'s by $\bar{\Gamma}$. ($\bar{\Gamma}$ is usually not the collection of all subsets of C , for which we have already introduced the notation 2^C). A set containing a finite (or zero) number of points is a finite set. ($\emptyset$ is a finite set.) A finite subset of C will be denoted by γ , and the collection of all finite subsets of C will be denoted by Γ . We apply this correspondance between the different alphabets to other letters.

We note that every system of subsets is reflexive and properly ordered.

3.2 Lemma. Every reflexive and properly ordered system is isomorphic to a system of subsets of some set.

Proof. Let $\mathcal{C} = (C,>)$ be a reflexive and properly ordered system. To c' we let correspond $\bar{\gamma}(c') = \{c \mid c < c'\}$. The reader may easily show that this is a one-to-one correspondance between C and a collection $\bar{\Gamma}$ of subsets of C (here $\bar{\Gamma}$ will never consist of all the subsets of C), and that this correspondance takes $>$ into $\supset$ and $\supset$ into $>$.

It is well known that an equivalence relation (that is, one which generates a reflexive, transitive and symmetric ordered system) divide the set on which it is defined into mutually exclusive equivalence classes. If we denote the equivalence relation by $\sim$ and the equivalence class containing a by $[a]$, then $b \in [a]$ if and only if $b \sim a$. We remark that every ordered system has a natural equivalence relation defined by

3:3 $a' \sim a''$ **if either** $a' = a''$ **or both** $a' > a''$ **and** $a'' < a'$.

We may define an ordering relation on the system of equivalence classes by letting $[a] > [b]$ if $a > b$. The reader can easily show that this definition is independent of the particular representatives a and b of the classes $[a]$ and $[b]$.

3.4 **. Such an ordered system of equivalence classes is properly ordered.**

We use $[\mathcal{A}]$ for the properly ordered system derived from $\mathcal{A}$ in this way.

3.5 **. If** $\mathcal{A}$ **is reflexive, then so is** $[\mathcal{A}]$ **, and conversely.**

So far, except for systems of subsets and equivalence relations, we have used $>$ as the symbol of the ordering relation in which we were interested. There is, of course, no necessity for this, and we will use several other symbols for particular relations. However, the usage of terms we are about to discuss is usually restricted to ordered systems whose ordering relation is written $>$ or $\geq$.

If $a > b$ for all $b \in B$, then a is an upper bound of B . We define lower bound similarly. Any of the upper bounds of B which is a lower bound of the class of all upper bounds of B is a supremum (sometimes called "least upper bound" or "join") of B . A given B may have no supremum, one supremum, or several suprema. When we say " a is a supremum of B ", we should, to be precise, say " a is a supremum in $\mathcal{A}$ of B ", but it will usually be clear what ordered system is meant. If $\mathcal{A}$ is properly ordered, then $B \subset \mathcal{A}$ has at most one supremum. An infimum ("greatest lower bound" or "meet") is defined similarly and is subject to simular remarks. We use the symbols $\vee$ and $\wedge$ for suprema and infima. Thus, $a' \vee a''$ is a supremum of a' and a''). In particular cases (for an example see V-2), where the system is not properly ordered, we may use some definite way to define $a' \vee a''$ as a particular supremum of a' and a'' . We denote the supremum of $\{b_t | T\}$ variously as: $\vee_t b_t$, $\vee\{b_t | T\}$, $\vee_t \{b_t | T\}$. We use $\wedge$ for the infimum in a similar way. We depart from these notations if the elements are real numbers and the relation is "greater than or equal to" in its elementary sense, when we use "Sup" and "Inf" instead of $\vee$ and $\wedge$.

If each pair of elements (not necessarily distinct) has at least one supremum and at least one infimum (the pair a , a need have a neither as a supremum nor as an infimum), the ordered system is a trellis. If every subsystem has at least one supremum and one infimum, then the trellis is a complete trellis. The reader familiar with the definition of a lattice will see that a lattice is a reflexive and properly ordered trellis. Hence (3.5) the ordered system of the equivalence classes of a reflexive trellis is a lattice.

We suggest the consideration of the following ordered systems (and, possibly, some of their naturally arising subsystems); they should serve to make the meaning of these definitions clear.

1) The elements are the positive integers, or the positive and negative integers, with one of the relations: $>$ (definitely greater than), $\geq$ (greater than or equal to), $\backslash$ (divides), $\equiv$ (mod p) (congruent modulo p).

2) The elements are all finite sets of real numbers with one of the following relations: $>'$ meaning "contains more points than" , $>''$ meaning "contains at least as many points as" , $>'''$ meaning "contains exactly as many points as" .

3) The elements are all measurable subsets of the closed interval $[0,1]$, or of the line $(-\infty,+\infty)$, with $A >^{o} B$ meaning $\text{meas}(B-A) = 0$.

4) The elements are a , b , and c , and the relation holds in the following cases only: $a > a$, $b > a$, $c > a$, $c > b$, $c > c$.

5) Query: Is any trivially ordered system a trellis?

4. Functions. We have no occasion to deal with functions which are not single-valued. Hence "function" will be understood to mean "single-valued function" . We shall indicate the domain of definition of a function in the notation for the function. A function defined on the points of X is, for example, $f(x|X)$. $f(x)$ is the value of f at x ; we never use $f(x)$ to refer to a function. With certain exceptions, such as f as a function defined on X and having values in Y , the

letter denoting the function is often associated with the set in which its values lie. Thus $a(b|B)$ is a function defined for b in B, and is likely to have its values in A, or perhaps in A'' or A_d.

We remind the reader of a few simple facts about such functions. If f is defined for x in X and has values in Y, then there is a function defined for H in 2^X and having its values in 2^Y, which is also denoted by f and is defined by

$$f(H) = \{y | y = f(x),\ x \in H\}.$$

This process can be, and is, continued to define a function from 2^{2^X} to 2^{2^Y}, and so on. The function f^{-1} from 2^Y to 2^X is defined by

$$f^{-1}(K) = \{x | f(x) \in K\}.$$

Again, we may repeat this process to define a function from 2^{2^Y} to 2^{2^X}, and so on. We make this transfer from point-function to set-function or collection-function freely at all times. We have $f(f^{-1}(H)) = H$, but only $f^{-1}(f(H)) \supset H$.

$f(a|A)$ is an extension of $g(b|B)$ if $B \subset A$ and $f(b) = g(b)$ for each $b \in B$.

The notation distinguishes between $f(g(x|X))$ and $f(g(x)|X)$. When we write $f(g(x|X))$ we imply that g is a function defined on X, and that, as a point, $g(x|X)$ belongs to the domain of definition of f, whose value at $g(x|X)$ is $f(g(x|X))$. When we write $f(g(\underline{X})|X)$ we imply that g is a function defined (at least) on X, and that $g(X)$ is a part of the domain of definition of f; in these circumstances, $f(g(X)|X)$ is a function defined on X, whose value at x is $f(g(x))$.

<u>5. Products</u>. The formulation of the product of ordered sets that we now give may seem strange at first, but I feel that it is the natural and convenient one. The product of two simple sequences is usually described as the set of pairs (m', m'') where m' and m'' are positive integers and $(m', m'') > (n', n'')$ if both $m' > n'$ and $m'' > n''$. We prefer to regard the elements of the product as functions defined for the points 1 and 2 with functional values in a simple sequence. This is the notion that we are going to generalize.

If, for each $c \in C$, $\mathfrak{a}^c = (A^c, >^c)$ is an ordered system, and if Γ is a system of subsets of C, then we consider all the functions $a(c|\bar{\gamma})$, where $\bar{\gamma} \in \bar{\Gamma}$ and $a(c) \in A^c$ for all $c \in \gamma$. We define $>$ by: $a'(c|\gamma') > a''(c|\gamma'')$ if for each $c \in \gamma''$ we have $c \in \gamma'$ (hence $\gamma' \supset \gamma''$) and $a'(c) >^c a''(c)$. The resulting ordered system of functions is called the <u>product</u> of the

$\mathfrak{a}^c$ by Γ, and is written $(\Gamma, \mathfrak{a}^c | C)$. The systems $\mathfrak{a}^c$ are factors of $(\Gamma, \mathfrak{a}^c | C)$. If there is no $\gamma \in \Gamma$ such that $c \in \gamma$ then $\mathfrak{a}^c$ is an inessential factor.

In the example above, $C = \{1,2\}$, Γ consists of a single subset of C, namely C itself, and $\mathfrak{a}^1$ and $\mathfrak{a}^2$ are each the sequence of positive integers in their natural order. If Γ consists of C alone, then we write $P_C \mathfrak{a}^c$ $P\{\mathfrak{a}^c | C\}$, or $P_C \{\mathfrak{a}^c | C\}$ for $(\Gamma, \mathfrak{a}^c | C)$. We call such a product the product of the $\mathfrak{a}^c$. This is, of course, the classical case. Thus the example above is the product of two simple sequences.

A product of ordered systems is ordered. A product is properly ordered, vacuously ordered, reflexive, or irreflexive, if all its essential factors have the corresponding property.

If $\{A^c | C\}$ is a collection of sets, then we may regard the sets as vacuously ordered systems and form $P\{\mathfrak{a}^c | C\}$. This product is a vacuously ordered system and may be considered a set. This set is called the product of the sets A .

There is a slight further generalization of this notion of product, which we shall not need, and which we shall omit.

6. Zorn's Lemma. We take an important and convenient lemma due to Max Zorn as an axiom. It is equivalent to the axiom of choice, and is a more convenient and powerful tool than transfinite induction. It also has the advantage that it can be easily stated in elementary terms.

An element a' of the ordered system (A, >) is maximal if a>a' implies a'>a . (This coincides with the usual definition for a properly and linearly ordered system.)

6.1 Zorn's Lemma (First Form). An ordered system, each of whose linear subsystems has an upper bound, contains a maximal element.

6.2 Zorn's Lemma (Second Form). Every ordered system contains a linear subsystem B, such that every upper bound of B (if any exist) belongs to B.

A property of sets is of finite character, if a set has the property when and only when all its finite subsets have the property.

6.3 Zorn's Lemma (Third Form). Given a set and a property of finite character, there exists a maximal subset having the property.

A condition on a function is a finite restriction, if, as a property of the graph of the function, it is of finite character. This means that it is the logical sum of conditions each of which depends on the functional values at a finite number of points. Some examples of finite restrictions are: that the function--

1) be a constant.
2) vanish on a certain set.
3) be additive (if defined on a group, linear space, etc.)
4) be bounded in absolute value by a particular function.

6.4 Zorn's Lemma (Fourth Form). The class of those functions defined on subsets of a given set and satisfying a given family of finite restrictions, contains a function no one of whose extentions belongs to the class.

We suggest some results and methods of proof; the reader is advised to complete the proofs.

1) 6.1 implies that, if A and B are sets, there is either a one-to-one mapping of A onto a subset of B or a one-to-one mapping of B onto a subset of A . (That is, the cardinal numbers of A and B are comparable.) For the relation of extension orders the system of all one-to-one mappings of a subset of A onto a subset of B .

2) 6.3 implies that any set of numbers contains a maximal set algebraically independent over a given field. For the property of being algebraically independent over a given field is of finite character.

3) 6.4 implies the Hahn-Banach theorem (Banach 1932, pp. 28-29). For the properties of being additive, homogeneous, and bounded by p , are finite restrictions.

4) 6.1 implies 6.2. For the system of all linear subsystems of a given system, when ordered by inclusion, satisfies the hypothesis of 6.1.

5) 6.1 implies 6.3. For the system of subsets of the given set having the given property of finite character, when ordered by inclusion, satisfies the hypothesis of 6.1.

6) 6.2 implies 6.1. For an upper bound for the linear subsystem of 6.2 is clearly maximal.

7) 6.3 implies 6.2. For the property of being a linear subsystem is a property of finite character.

8) 6.3 implies 6.4. For consider the graphs of the functions.

9) 6.4 implies 6.3. For the property of being a characteristic function is a finite restriction, and a property of finite character for subsets is a finite restriction for the corresponding characteristic functions.

7. Cardinal Numbers. We need a very little of the theory of cardinal numbers, which we now recall. Two sets have the same cardinal number (or potency), if there exists a one-to-one correspondence between them. (Compare 1) above.) The cardinal number of A is denoted by $|A|$. We write $|A| \leq |B|$ if there is a one-to-one mapping of A on a subset of B .

A set is countably infinite if it has the same cardinal number ($\aleph_0$) as the set of positive integers. A set is countable if it is finite or countably infinite. The union of a countable family of countable sets is countable. The product of a finite collection of countable sets is countable. Any subset of a countable set is countable.

Chapter II

DIRECTION

1. Introduction.
2. Directed Systems.
3. An Ordering.
4. Systems of Subsets.
5. Stacks.
6. The Countable Case.
7. The General Case.

1. Introduction. In this chapter we are concerned with the theory of directed systems for its topological applications. We discuss, therefore, only a part of the theory; a fuller account will appear soon in another place.

2. Directed Systems. A directed system is a non-empty ordered system, $\mathcal{A} = (A, >)$, in which we have

2.1 For every a' and a", and some a*, a*>a', a*>a".

This property was originally called "the composition property" (Moore and Smith 1922). We note that 2.1 may hold for $(A,>)$ and not for $(A,<)$. In this and the following chapters, when we write $\mathcal{A}$ we assume, unless the contrary is specified, that $\mathcal{A} = (A,>)$ and that $\mathcal{A}$ is directed. The same convention applies to $\mathcal{A}' = (A',>)$, $\mathcal{B} = (B,>)$, etc. We often use $\mathcal{A}$ where A should strictly be used. as $a \in \mathcal{A}$ for $a \in A$.

Three characteristic examples of directed systems are:
1° Any linearly ordered set.
2° A system of subsets of some set, directed by $\supset$; for example, the system of all finite subsets of a set.
3° The neighborhoods of a given point in a space, which are directed by $\subset$.

There are two important classes of subsystems of a directed system; the cofinal subsystems and the residual subsystems. A subsystem $\mathcal{B}$ of a directed system $\mathcal{A}$ is cofinal in $\mathcal{A}$ if

2.2 For each a, and some b, b>a.

$\mathcal{B}$ is residual in $\mathcal{A}$ if

2.3 For some a', and all a>a', $a \in \mathfrak{B}$.

These concepts are related by

2.4 **. $\mathfrak{B}$ is cofinal in $\mathfrak{A}$ if and only if $\mathfrak{A}-\mathfrak{B}$ is not residual in $\mathfrak{A}$. Hence $\mathfrak{B}$ and $\mathfrak{A}-\mathfrak{B}$ cannot both fail to be cofinal in $\mathfrak{A}$.**

2.5 **. Every residual subsystem of $\mathfrak{A}$ is cofinal in $\mathfrak{A}$.**

It is important that we have

2.6 **. If $\mathfrak{A}$ is a directed system and $\mathfrak{B}$ is cofinal in $\mathfrak{A}$, then $\mathfrak{B}$ is a directed system.**

We observe that a subsystem of a directed system, which surely is ordered, need not be directed. Clearly,

2.7 **. If $\mathfrak{C}$ is cofinal in $\mathfrak{B}$ and $\mathfrak{B}$ is cofinal in $\mathfrak{A}$, then $\mathfrak{C}$ is cofinal in $\mathfrak{A}$.**

2.8 **. If $B' \supset B''$, and $\mathfrak{B}''$ is cofinal in $\mathfrak{A}$, then $\mathfrak{B}'$ is cofinal in $\mathfrak{A}$.**

A most important notion is that of cofinal similarity. $\mathfrak{A}$ and $\mathfrak{B}$ are <u>cofinally similar</u> if there exist $\mathfrak{A}'$, $\mathfrak{B}'$, $\mathfrak{C}'$ so that

2.9 $\mathfrak{A} \cong \mathfrak{A}'$, $\mathfrak{A}'$ and $\mathfrak{B}'$ are both cofinal in $\mathfrak{C}'$, $\mathfrak{B}' \cong \mathfrak{B}$.

If this is so, we write $\mathfrak{A} \sim \mathfrak{B}$

2.10 Lemma. If $\mathfrak{C}$ is cofinal in both $\mathfrak{A}$ and $\mathfrak{B}$, and $\mathfrak{A} \cap \mathfrak{B} = \mathfrak{C}$, then we may define > on $D = A \cup B$ so that both $\mathfrak{A}$ and $\mathfrak{B}$ are cofinal in $\mathfrak{D}$.

<u>Proof</u>. Initially, $d' > d''$ is defined (to hold or not to hold) if d' and d'' either both belong to $\mathfrak{A}$ or both belong to $\mathfrak{B}$. We extend the definition of $>$ by defining

$$a > b \text{ , if, for some } c \text{ , } a > c > b \text{ ;}$$
$$b > a \text{ , if, for some } c \text{ , } b > c > a \text{ .}$$

We leave to the reader the verification that $>$ now orders and directs $\mathfrak{B}$ and that $\mathfrak{A}$ and $\mathfrak{B}$ are cofinal in $\mathfrak{D}$

2.11 Theorem. $\sim$ is an equivalence relation.

<u>Proof</u>. It is clear that $\sim$ is reflexive and symmetric; we need only show that it is transitive. Let $\mathfrak{A} \sim \mathfrak{B} \sim \mathfrak{C}$; then there exist $\mathfrak{A}'$, $\mathfrak{B}'$, $\mathfrak{D}'$, $\mathfrak{B}''$, $\mathfrak{D}''$, $\mathfrak{C}''$ so that $\mathfrak{A} \cong \mathfrak{A}'$, $\mathfrak{A}'$ and $\mathfrak{B}'$ are both cofinal in $\mathfrak{D}'$, $\mathfrak{B}' \cong \mathfrak{B} \cong \mathfrak{B}''$, $\mathfrak{B}''$ and $\mathfrak{C}''$ are both cofinal in $\mathfrak{D}''$, $\mathfrak{C}'' \cong \mathfrak{C}$. Clearly, by changing some of the primed directed systems to isomorphic systems, we may, without disturbing these relations, obtain $\mathfrak{B}' = \mathfrak{B}''$ and $D' \cap D'' = B' = B''$. We may then (2.10) direct $E = D' \cup D''$ so that $\mathfrak{D}'$ and $\mathfrak{D}''$ are cofinal in $\mathfrak{E}$; hence

(2.7) $\mathcal{A}'$ and C'' are both cofinal in $\mathcal{E}$. Since $\mathcal{A}\cong\mathcal{A}'$ and $C''\cong C$, we have $\mathcal{A}\sim C$ and the transitivity is proved.

As an equivalence relation, $\sim$ divides the family of directed systems into classes called cofinal types.

3. An ordering. We say that $\mathcal{A} > \mathcal{B}$ if there exist functions $a(b|B)$ and $b(a|A)$ such that

3.1 $a>a(b)$ implies $b(a)>b$.

We have

3.2 **. $\mathcal{A}>\mathcal{B}$ and $\mathcal{B}>C$ implies $\mathcal{A}>C$; hence $>$ orders any family of directed systems.**

3.3 **. $\mathcal{A}>\mathcal{B}$ and $\mathcal{B}>\mathcal{A}$ if and only if there exist functions $a(b|B)$ and $b(a|A)$ such that, $a>a(b)$ implies $b(a)>b$, and $b>b(a)$ implies $a(b)>b$.**

The reader interested in the meaning of 3.1 is advised to consider the following examples of directed systems and decide which ordered pairs satisfy 3.1.

1° The integers in their natural order.
2° The system of all finite sets of integers.
3° The system of all finite sets of real numbers.
4° The system of all countable sets of integers.
5° The system of all countable sets of real numbers.

3.4 Theorem. $\mathcal{A}>\mathcal{B}$ and $\mathcal{B}>\mathcal{A}$ if and only if $\mathcal{A}\sim\mathcal{B}$.

Proof. Suppose $\mathcal{A}\sim\mathcal{B}$, then, for some $\mathcal{A}'$, $\mathcal{B}'$ and C', we have $\mathcal{A}\cong\mathcal{A}'$, both $\mathcal{A}'$ and $\mathcal{B}'$ are cofinal in C', $\mathcal{B}'\cong\mathcal{B}$. It is clear that $\mathcal{A}>\mathcal{A}'$, $\mathcal{A}'>\mathcal{A}$, $\mathcal{B}'>\mathcal{B}$, $\mathcal{B}>\mathcal{B}'$. If $a'\in\mathcal{A}'$, then, since $\mathcal{B}'$ is cofinal in C', we may choose $b'\in\mathcal{B}'$ so that $b'>a'$. Hence we may choose $b'(a'|A')$ so that $b'(a')>a'$ for each a'; similarly we may choose $a'(b'|B')$ so that $a'(b')>b'$ for each b'. Hence (3.3) $\mathcal{A}'>\mathcal{B}'$ and $\mathcal{B}'>\mathcal{A}'$, so that (3.2) $\mathcal{A}>\mathcal{B}$ and $\mathcal{B}>\mathcal{A}$.

Suppose $\mathcal{A}>\mathcal{B}$ and $\mathcal{B}>\mathcal{A}$. We assume that $A\cap B=\emptyset$, which can be obtained by considering a system isomorphic to $\mathcal{B}$. There exist (3.3) $a(b|B)$ and $b(a|A)$ such that, $a>a(b)$ implies $b(a)>b$, and $b>b(a)$ implies $a(b)>a$. We now extend the definition of $>$ to $D=A\cup B$ by setting

$$a>b\ ,\quad \text{if}\quad a>a(b)\ ;$$
$$b>a\ ,\quad \text{if}\quad b>b(a)\ .$$

It is clear that $\mathcal{D}$ is directed by $>$ and that $\mathcal{A}$ and $\mathcal{B}$ are cofinal in $\mathcal{D}$. Hence $\mathcal{A}\sim\mathcal{B}$.

3.5 **. $\mathcal{A}\sim\mathcal{B}$ if and only if there exist functions $a(b|B)$ and $b(a|A)$ such that, $a>a(b)$ implies $b(a)>b$, and $b>b(a)$ implies $a(b)>a$.**

The interested reader might prove

3.6 **. If C is finite, then $P\{A^c|C\}$ is a supremum of $\{A^c|C\}$.**

While we shall not need this result, it has considerable intrinsic interest, since it shows that $>$ directs the class of "all" directed systems.

4. Systems of Subsets. We shall see that each cofinal type contains at least one system of subsets. We use the methods of I-3.

4.1 Lemma. If $\mathfrak{A}$ is directed, then so is $[\mathfrak{A}]$, and these two systems are cofinally similar.

Proof. We apply 3.5, where (setting $\mathfrak{B} = [\mathfrak{A}]$) $a(b)$ is some representative of $b = [a]$, and $b(a) = [a]$. We leave the details to the reader.

4.2 Lemma. If $\mathfrak{A}=(A,>)$ is directed and if $\mathfrak{A}'=(A,\geq)$, where $a'\geq a''$ if either $a'>a''$ or $a'=a''$; then $\mathfrak{A}'$ is directed, and $\mathfrak{A}$ and $\mathfrak{A}'$ are cofinally similar

Proof. Let $a'(a|A)$ be such that $a'(a) > a$, let $\mathfrak{B} = \mathfrak{A}'$, and let $b(a) = a'(a)$ and $a(b) = a'(a)$, where $a = b$; then the lemma follows from 3.5.

4.3 **. Two isomorphic directed systems are cofinally similar.**

From 4.1, 4.2, 4.3, I-3.2 and I-3.5 we see immediately that

4.4 **. Each cofinal type contains at least one system of subsets.**

For many interesting problems, which lie outside the scope of the present discussion, we could restrict ourselves to those systems of subsets which are ideals in the ring of all subsets of a set.

5. Stacks. A stack (this term is due to M. M. Day) is the system of all finite subsets of some set, called the base of the stack. We recall that Γ denotes the stack with base C, etc.

The topological importance of stacks is related to

5.1 Theorem. For any directed system $\mathfrak{D}$, we have $\Delta>\mathfrak{D}$.

Proof. Let $\delta\in\Delta$, then since δ is finite, there exists a $d(\delta)\in\mathfrak{D}$, such that $d(\delta) > d$ for all $d\in\delta$. If we let $\delta(d) = d$ (the set made up of d alone), then $\delta\supset\delta(d)$ means $d\in\delta$ which implies $d(\delta) > d$. Hence 3.1 holds, and $\Delta>\mathfrak{D}$.

6. The Countable Case. Those directed systems which have at most a countable number of elements are of interest. Clearly,

every directed system containing a maximal element is cofinally similar to every other such system. If $\mathcal{a}$ is a directed system without maximal element containing at most a countable number of elements, then we may suppose that $\{a_n | N\}$ is an enumeration of the system. Set $b_1 = a_1$, and choose b_n so that $b_n > a_n$ and $b_n > b_{n-1}$. The subsystem $\mathcal{B}$ is obviously cofinal in $\mathcal{a}$. If, for some n', we had $b_n < b_{n'}$ for all $n > n'$, then $b_{n'}$ would be a maximal element of $\mathcal{B}$ and hence of $\mathcal{a}$, contrary to hypothesis. Since this cannot happen, we can choose a subsystem $C = \{c_k = b_{nk} | K\}$ of $\mathcal{B}$, such that $c_k > c_{k-1}$ and $c_k \not< c_{k-Z}$ for all k. It is clear that C is cofinal in $\mathcal{B}$ and hence in $\mathcal{a}$ and it is also clear that C is isomorphic to the system of positive integers in their natural order, that C is, in fact, a simple sequence. Hence we have proved

6.1 Theorem. A directed system with a countable number of elements contains either a maximal (that is, cofinal) element or a cofinal simple sequence. In particular, every such system belongs to one of two cofinal types, the one containing the finite directed system or the one containing the simple sequence.

7. The General Case. The simplicity of the countable case is absent in the general case. Tho there remain many unsolved problems, we can give some idea of the situation. We shall omit all semblance of proof.

There is a family of easily accesible cofinal types. The structure of the lower part of this family with regard to $>$ is

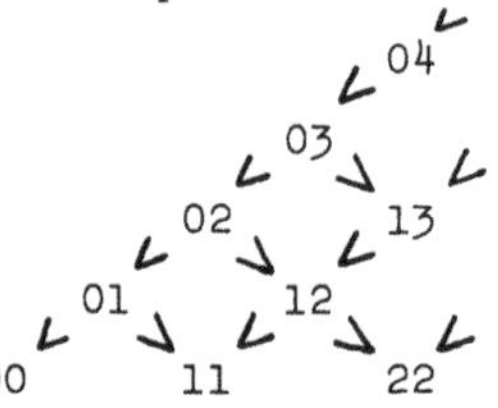

Here the indicated relations generate all those which exist by transitivity (thus $02 > 12 > 11$ implies that $02 > 11$; and it is not true that $01 > 22$). All these cofinal types are $>$ the type containing all finite directed systems.

The types 00 , 01 , 02 , etc. contain stacks. The types 11 , 22 , 33 , etc. contain transfinite sequences. We see that "transfinite sequence > stack" never occurs. The cofinal similarity between simple sequence and stack on a countable base is purely the result of countability and does not generalize.

From this family we can generate a complete lattice of cofinal types. However it is not known whether the system of all cofinal types (suitably restricted to avoid paradoxes) is a complete lattice or not. Neither is it known whether each pair of cofinal types has an infimum or not.

Chapter III

CONVERGENCE

1. Introduction.
2. Convergence and Phalanxes.
3. The Bill of Rights.
4. Effectiveness.
5. Relativization.
6. Open Sets and T-spaces.
7. Continuity.
8. Separation Axioms.
9. Historical Remarks.

1. Introduction. The aim of this chapter is twofold: to establish convergence as a basic concept equivalent, in a very wide class of spaces, to closure and neighborhoods; and to make convergence available as a tool in spaces where the basic concept is closure or neighborhoods. We have also collected here some other topological results of an elementary nature which we will need later.

In §2 we discuss the applicability of the term "convergence" and point out the phalanx as particularly useful. In §3 we establish the equivalence of closure, convergence and neighborhoods under very general conditions. In §4 we discuss the effectiveness of the various directed systems as carriers of convergence. In §§5 and 6 we discuss relativization and open sets. In §§7 and 8 we deal briefly with continuity and the separation axioms. In §9 we make some historical remarks.

The results of this chapter may be summarized as follows: The term "convergent" concerns functions defined on directed systems. Convergence is equivalent to closure and neighborhoods in all reasonable spaces. When we use convergence in a general problem we need consider only phalanxes.

2. Convergence and Phalanxes. While the notion of convergence can be further generalized (cp. Garrett Birkhoff 1939), I feel that, for the main purposes of topology, the convenient class of objects, whose "convergence" is to be considered, is the class of functions defined on a directed system and taking values in a space. All the further generalizations that I know which

preserve uniqueness of limit in the classical spaces can be reduced to this notion. This was the attitude taken by Moore and Smith in 1922 when they discussed the convergence of real-valued objects.

Classically, one considered sequences; a sequence is precisely a function on the directed system of the integers.

Stacks (II-5) play a special role in the theory of directed systems; functions defined on stacks play a special role in problems of convergence. We call such a function a phalanx. The base of the stack is the base of the phalanx. A phalanx with base A is an A-phalanx.

We shall modify our functional notation when it is convenient. A function on the directed system $\mathcal{A}$ will be denoted by $x(a|\mathcal{A})$ as well as $x(a|A)$ in order to refer to the ordering relation of $\mathcal{A}$. If $\mathcal{A}$ is the stack Γ, we write $x(\gamma|C)$ rather than $x(\gamma|\Gamma)$.

One reason for the importance of the phalanx is its ease of manipulation. A phalanx $x(\alpha|A)$ is inflated to a phalanx $x_1(\alpha\cup\beta\,|\,A\cup B)$ by defining $x_1(\alpha\cup\beta) = x(\alpha)$. (Here A and B are disjoint.) This essentially introduces dummy elements and is important for results about the selection of suitable subphalanxes. Two phalanxes $x(\alpha|A)$ and $y(\beta|B)$ are meshed to a phalanx $z(\gamma|C)$, where $C = A\cup B$, when

$$z(\gamma) = \begin{cases} x(\alpha|A), & \text{if } |\gamma| \text{ is odd,} \\ y(\beta|B), & \text{if } |\gamma| \text{ is even.} \end{cases}$$

3. The Bill of Rights. We now set forth the relations between closure, convergence and neighborhoods in very general spaces. We are dealing with a certain set X, its subsets and points. A closure operator assigns to each subset H of X a closure $\bar{H}$. We shall usually assume

3.1 $\bar{\emptyset} = \emptyset$, $\overline{H\cup K} = \bar{H}\cup\bar{K}$,

of which an easy consequence is

3.1a $H\subset K$ **implies** $\bar{H}\subset\bar{K}$

(but we shall not assume, for the present, more special relations, such as $\bar{H} = \bar{\bar{H}}$, $\bar{H} = H$). A notion of convergence determines, for each function $x(a|\mathcal{A})$ from a directed system to X, whether $x(a|\mathcal{A})$ converges to x or not. The statement " $x(a|\mathcal{A})$ converges to x " is also written $x(a)\xrightarrow[\mathcal{A}]{} x$, or merely $x(a)\longrightarrow x$; another expression of this statement is " x is a limit of $x(a|\mathcal{A})$ ". Similarly, " $x(\gamma|C)$ converges to x " is also written $x(\gamma)\xrightarrow[C]{} x$, or merely $x(\gamma)\longrightarrow x$. We usually assume

3.2 If $x(a|\mathfrak{A})$ converges to x, and $\mathfrak{B}$ is cofinal in $\mathfrak{A}$, then $x(b|\mathfrak{B})$ also converges to x.

A neighborhood system (usually abbreviated nbd system) assigns to each point x of X a non-empty family of subsets of X called the neighborhoods (nbds) of x. We usually assume

3.3 If N_1 and N_2 are nbds of x, then some $N_3 \subset N_1 \cap N_2$ is a nbd of x. (The nbds of x are directed by $\subset$)

These notions are often connected by

3.4 $x(a|\mathfrak{A})$ converges to $x \Longleftrightarrow$ whenever $\mathfrak{A}'$ is cofinal in $\mathfrak{A}$ $x \in \overline{x(\mathfrak{A}')}$.

3.5 N is a nbd of $x \Longleftrightarrow x \quad \overline{X-N}$.

3.6 N is a nbd of $x \Longleftrightarrow$ whenever $x(a|\mathfrak{A})$ converges to x, $N \cap x(\mathfrak{A}) \neq \emptyset$.

3.7 $x \in \overline{H} \Longleftrightarrow$ for some $\mathfrak{A}$ and $x(a|\mathfrak{A})$, $x(a|\mathfrak{A})$ converges to x and $x(\mathfrak{A}) \subset H$.

3.8 $x \in \overline{H} \Longleftrightarrow$ for each nbd N of x. $N \cap H \neq \emptyset$.

3.9 $x(a|\mathfrak{A})$ converges to $x \Longleftrightarrow$ for each nbd N of x, and some $a' \in \mathfrak{A}$, $a > a'$ implies $x(a) \in N$.

These may be modified to

3.6p N is a nbd of $x \Longleftrightarrow$ whenever (for any C) $x(\gamma|C)$ converges to x, $N \cap x(\Gamma) \neq \emptyset$.

3.7p $x \in \overline{H} \Longleftrightarrow$ for some C and some $x(\gamma|C)$ converging to x, $x(\Gamma) \subset H$.

Further modification yields, where $D = 2^X$,

3.6pp N is a nbd of $x \Longleftrightarrow$ whenever $x(\delta|D)$ converges to x, $N \cap x(\Delta) \neq \emptyset$.

3.7pp $x \in \overline{H} \Longleftrightarrow$ some $x(\delta|D)$ converges to x, with $x(\Delta) \subset H$.

3.10 $\overline{\emptyset} = \emptyset$, $\overline{H \cup K} = \overline{H} \cup \overline{K}$.

3.11 $x(a|\mathfrak{A})$ converges to $x \Longleftrightarrow$ for each $\mathfrak{A}'$ cofinal in $\mathfrak{A}$ and some $x'(\delta|D)$ converging to x, $x'(\Delta) \subset x(\mathfrak{A}')$.

3.12 If $M \supset N$, and N is a nbd of x, then M is a nbd of x. If N_1 N_2 are nbds of x, then $N_1 \cap N_2$ is a nbd of x.

We have,

3.13 **. 3.10 implies 3.1; 3.11 implies 3.2; 3.12 implies 3.3.**

In view of this result, we call 3.1, 3.2 and 3.3 the weak conditions and 3.10, 3.11, and 3.12 the strong conditions. One notion is derivable from another if the proper one of 3.4 thru 3.9 holds. Thus, if 3.7 holds, the closure is derivable from the convergence. Several notions are mutually derivable if each is derivable from each other. We have,

3.14 Theorem. The notions of closure, convergence and nbds are equal in that

1) If a notion satisfies a weak condition, then the two notions derivable from it satisfy the strong conditions and are mutually derivable.

2) If a notion satisfies the strong condition, then the two notions derivable from it satisfy the strong conditions, and all three notions are mutually derivable.

Furthermore, if the convergence satisfies the strong condition, then we may replace 3.6 by 3.6p or 3.6pp and 3.7 by 3.7p or 3.7pp in the definition of derivability.

Proof. The proof of this theorem naturally falls into several parts.

1°. 3.1, 3.4 and 3.5 imply 3.12, 3.11, 3.6, 3.6p, 3.6pp, and 3.9.

We start with a closure satisfying 3.1. If $M \supset N$ and N is a nbd of x, then (3.5) $x \notin \overline{X-N}$. Since $X-M \subset X-N$, we have (3.1a) $\overline{X-M} \subset \overline{X-N}$; hence $x \notin \overline{X-M}$, and (3.5) M is a nbd of x. If N_1 and N_2 are nbds of x, then (3.5) $x \notin \overline{X-N_1}$, $x \notin \overline{X-N_2}$. Since $X-(N_1 \cap N_2) = (X-N_1) \cup (X-N_2)$, we have (3.1) $\overline{X-(N_1 \cap N_2)} = \overline{X-N_1} \cup \overline{X-N_2}$; hence $x \notin \overline{X-(N_1 \cap N_2)}$, and (3.5) $N_1 \cap N_2$ is a nbd of x. Hence 3.12 holds.

Now if $x(a|\mathcal{A})$ converges to x, then (3.4), if $\mathcal{A}'$ is cofinal in $\mathcal{A}$ we have $x \in \overline{x(\mathcal{A}')}$. Let $H = X - x(\mathcal{A}')$; then (3.5) H is not a nbd of x, and, in fact, (3.12) H contains no nbd of x. Let $\delta \in \Delta$, then δ is a finite collection of subsets of X. Some of these sets may be nbds of x; let these be $\{K\}$. Now (3.12) $\cap\{K\}$ is a nbd of x, hence is not contained in H. Hence we may choose $x_1(\delta) \in \cap\{K\}$, so that $x_1(\delta) \notin H$. (If $\{K\}$ is empty, as surely will happen when $\delta = \emptyset$, then (I-2) $\cap\{K\}$ is X, and no difficulty arises.) Now $x_1(\delta) \notin H$ means $x_1(\delta) \in x(\mathcal{A}')$. If $x_1(\delta)$ is chosen in this way for all $\delta \in \Delta$, and Δ' is cofinal in Δ, then I say that $x_1(\Delta') \cap N \neq \emptyset$. For N is an element of D, and if $\delta \in \Delta'$ is such that $N \in \delta$, then $x_1(\delta) \in N$. Now $x \in \overline{x_1(\Delta')}$; for if $x \notin \overline{x_1(\Delta')} = \overline{X-(X-x_1(\Delta'))}$, then (3.5) $X - x_1(\Delta')$ would be a nbd of x, whence $x_1(\Delta') \cap (X - x_1(\Delta')) \neq \emptyset$ which is a contradiction. Hence (3.4) $x_1(\delta|D)$ converges to x, and since each $x_1(\delta) \in x(\mathcal{A}')$ we have $x_1(\Delta) \subset x(\mathcal{A}'))$, and the first part of 3.11 is proved.

Now suppose that $x(a|\mathcal{A})$ and x are given, and that, for each $\mathcal{A}'$ cofinal in $\mathcal{A}$ there is an $x'(\delta|D)$ converging to x with $x'(\Delta) \subset x(\mathcal{A}')$. Now Δ is cofinal in Δ, hence (3.4) $x \in \overline{x(\Delta)}$, and, since (3.1a) $\overline{x'(\Delta)} \in \overline{x(\mathcal{A}')}$, we have

$x\in\overline{x(\mathcal{A}')}$. Hence (3.4) $x(a|\mathcal{A})$ converges to x, and the remainder of 3.11 is proved.

If N is a nbd of x, and $x(a|\mathcal{A})$ converges to x, then (3.4) $x\in\overline{x(\mathcal{A})}$; hence (as above) $X-x(\mathcal{A})$ contains no nbd of x. Therefore $X-x(\mathcal{A})\not\supset N$, that is, $N\cap x(\mathcal{A}) \neq \emptyset$. Thus one half of 3.6, 3.6p and 3.6pp is proved. The remainder of 3.6, 3.6p and 3.6pp follows from the argument used in proving the first half of 3.11.

Let $x(a|\mathcal{A})$ converge to x, and let N be a nbd of x. Let $\mathcal{B} = \{a|x(a)\notin N\}$, then (II-2.4) either $\mathcal{B}$ is cofinal in $\mathcal{A}$ or $\mathcal{A}-\mathcal{B}$ is residual in $\mathcal{A}$. If $\mathcal{B}$ were cofinal in $\mathcal{A}$, then (3.4) $x\in\overline{x(\mathcal{B})}$. But $x(\mathcal{B})\subset X-N$; hence, since (3.1a) $\overline{x(\mathcal{B})}\subset\overline{X-N}$, $x\in\overline{X-N}$ this contradicts (3.5) the hypothesis that N is a nbd of x. Hence $\mathcal{A}-\mathcal{B}$ is residual in $\mathcal{A}$; that is, there exists an a' such that $a > a'$ implies $a\notin\mathcal{B}$, which means $x(a)\in N$. One half of 3.9 is proved.

Suppose that if N is a nbd of x there exists an a' such that $a > a'$ implies $x(a)\in N$. Let $\mathcal{A}'\subset\mathcal{A}$ be such that $x\in\overline{x(\mathcal{A}')}$. Then (3.5) $X-x(\mathcal{A}')$ is a nbd of x and, for some a', $a > a'$ implies $x(a)\in X-x(A')$. Hence $\mathcal{A}'$ is not cofinal in $\mathcal{A}$. Hence if $\mathcal{A}'$ is cofinal in $\mathcal{A}$ we have $x\in x(\mathcal{A}')$ and (3.4) $x(a|\mathcal{A})$ converges to x. The remainder of 3.9 is proved.

2°. 3.2. 3.6 and 3.7 imply 3.10, 3.12, 3.5 and 3.8.

We begin with a convergence satisfying 3.2. Clearly $x(\mathcal{A})\subset\emptyset$ is impossible; hence (3.7) $x\in\overline{\emptyset}$ is impossible. Therefore $\overline{\emptyset} = \emptyset$. If $x\in\overline{H}$, then (3.7) some $x(a|\mathcal{A})$ converges to x, where $x(\mathcal{A})\subset H\subset H\cup K$; hence (3.7) $x\in\overline{H\cup K}$. Therefore $\overline{H}\cup\overline{K}\subset\overline{H\cup K}$. If $x\in\overline{H\cup K}$, then (3.7) some $x(a|\mathcal{A})$ converges to x and $x(\mathcal{A})\subset H\cup K$. Let $\mathcal{A}' = \{a|x(a)\in H\}$, $\mathcal{A}'' = \{a|x(a)\in K\}$; then $\mathcal{A}'\cup\mathcal{A}'' = \mathcal{A}$ and (II-2) either $\mathcal{A}'$ or $\mathcal{A}''$ is cofinal in $\mathcal{A}$. If $\mathcal{A}'$ is cofinal in $\mathcal{A}$, then (3.2) $x(a|\mathcal{A}')$ converges to x and (3.7) $x\in\overline{H}$. If $\mathcal{A}''$ is cofinal in $\mathcal{A}$, we may show in a similar way that $x\in\overline{K}$. Hence $\overline{H\cup K}\subset\overline{H}\cup\overline{K}$. Hence $\overline{H\cup K} = \overline{H}\cup\overline{K}$, and 3.10 is proved.

If N is a nbd of x, then (3.6) $N\cup x(\mathcal{A}) \neq \emptyset$, whenever $x(a|\mathcal{A})$ converges to x. If $M\supset N$, then surely $M\cap x(\mathcal{A}) \neq \emptyset$, whenever $x(a|\mathcal{A})$ converges to x. Hence (3.6) M is a nbd of x. If N' and N'' are such that $N'\cap N''\cap x(\mathcal{A}) = \emptyset$, where $x(a|\mathcal{A})$ converges to x, then, as above, either $\mathcal{A}'$ or $\mathcal{A}''$ is cofinal in $\mathcal{A}$, where $\mathcal{A}' = \{a|x(a)\in N'\}$, and $\mathcal{A}'' = \{a|x(a)\in N''\}$. If $\mathcal{A}'$ is cofinal in $\mathcal{A}$, then (3.2) $x(a|\mathcal{A}')$ converges to x ; hence (3.6), since $N'\cap x(\mathcal{A}') = \emptyset$, N' is not a nbd of x. Similarly, if $\mathcal{A}''$ is cofinal in , N'' is not a nbd of x. Hence, if N' and N'' are nbds of x, then $N'\cap N''\cap x(\mathcal{A}) \neq \emptyset$, whenever $x(a|\mathcal{A})$ converges to x.

Hence (3.6) $N' \cap N''$ is a nbd of x. So 3.12 holds.

If N is a nbd of x, then (3.6) $N \cap x(\mathfrak{A}) \neq \emptyset$, whenever $x(a|\mathfrak{A})$ converges to x. Hence we cannot have both $x(\mathfrak{A}) \subset X-N$ and $x(a|\mathfrak{A})$ converging to x; hence (3.7) $x \notin \overline{X-N}$. The remainder of 3.5 follows by reversing this argument.

If $x \in \overline{H}$, then (3.7) some $x(a|\mathfrak{A})$ converges to x with $x(\mathfrak{A}) \subset H$. If N is a nbd of x, then (3.6) $N \cap H \supset N \cap x(\mathfrak{A}) \neq \emptyset$. Conversely, if $H \cap N \neq \emptyset$ for all nbds of x, then, since $H \cap (X-H) = \emptyset$, $X-H$ is not a nbd of x, and (3.5) $x \in \overline{X-(X-H)} = \overline{H}$. Hence 3.8 holds.

3°. 3.3, 3.8 and 3.9 imply 3.10, 3.11, 3.4, 3.7, 3.7p and 3.7pp.

We begin with a nbd system satisfying 3.3. Any x has some nbd N; $\emptyset \cap N = \emptyset$; hence (3.8) $x \notin \overline{\emptyset}$. Hence $\overline{\emptyset} = \emptyset$. If $x \in \overline{H}$, then (3.8), for each nbd N of x, $(H \cup K) \cap N \supset H \cap N \neq \emptyset$. Hence (3.8) $x \in \overline{H \cup K}$. So we have $\overline{H} \cup \overline{K} \subset \overline{H \cup K}$. If $x \notin \overline{H} \cup \overline{K}$, then (3.8) for some nbds N' and N'' of x, $H \cap N' = \emptyset = K \cap N''$. Let $N^* \subset N' \cap N''$ be a nbd of x; then $(H \cup K) \cap N^* \subset (H \cup K) \cap (N' \cap N'') \subset (H \cap N') \cup (K \cap N'') = \emptyset$. Hence (3.8) $x \notin \overline{H \cup K}$. Hence $\overline{H \cup K} \subset \overline{H} \cup \overline{K}$. Hence $\overline{H} \cup \overline{K} = \overline{H \cup K}$. Thus 3.10 is proved.

Let $x(a|\mathfrak{A})$ converge to x. Let $\delta \in\!\in \Delta$ be a finite collection of subsets of x. Some of these, $\{N\}$, are nbds of x; then (3.9) there are a_N so that $a > a_N$ implies $x(a) \in N$. Since $\mathfrak{A}$ is directed we may choose $a \in a_N$ for this finite collection of a_N's. Let $x'(\delta) = x(a)$. Then, if $N \in \delta$ is a nbd of x, $N \in\!\in \delta$ implies $x'(\delta) \in N$. Hence (3.9) $x'(\delta|D)$ converges to x. Now 3.2 is an immediate consequence of 3.9. Hence we can apply the argument above to $x(a|\mathfrak{A}')$, where $\mathfrak{A}'$ is cofinal in $\mathfrak{A}$. Thus one half of 3.11 is proved.

If $x(a|\mathfrak{A})$ does not converge to x; then (3.9) for some N and every a there is an $a' > a$ such that $x(a') \notin N$. That is, $\mathfrak{A}' = \{a | x(a) \in N\}$ is cofinal in $\mathfrak{A}$. It is clear from this and 3.9 that the other half of 3.11 holds.

We proceed to the proof of 3.4. If $x(a|\mathfrak{A})$ converges to x, and $\mathfrak{A}'$ is cofinal in $\mathfrak{A}$, then (3.11, 3.13) $x(a|\mathfrak{A}')$ converges to x. Whence (3.9) $x(\mathfrak{A}') \cap N \neq \emptyset$, and (3.8) $x \in \overline{x(\mathfrak{A}')}$. To prove the converse of this result we proceed as in the proof of the latter half of 3.11, applying 3.8 as well as 3.9. This completes the proof of 3.4.

If $x \subset \overline{H}$, and if $\{N\}$ is a finite collection of nbds of x, then (3.3), for some nbd N' of x, $N \in\!\in \{N\}$ implies $N' \subset N$, and hence (3.8) $H \cap N' \neq \emptyset$. If $\delta \in\!\in \Delta$, and if $\{N\}$ consists of those nbds of x which belong to δ, then we may choose $x'(\delta) \in H$ so that $x'(\delta) \in\!\in \cap\{N\}$. Hence, as in the proof of 3.11, $x'(\delta|D)$ converges to x. Thus one half of 3.7, 3.7P and 3.7pp is proved. If $x(a|\mathcal{A})$ converges to x, and if $x(\mathcal{A})\subset H$, then (3.9), for each nbd N of x, $H\cap N\supset x(\mathcal{A})\cap N \neq \emptyset$; hence (3.8) $x\in\bar{H}$. This completes the proof of 3.7, 3.7p and 3.7pp.

4°. 3.10 and 3.4 imply 3.7, 3.7p and 3.7pp.

Let $x\in\bar{H}$. Let $\{K\}$ be the class of subsets of H such that $x\in\overline{H-K}$. If K' and K'' belong to $\{K\}$, then, since (3.10) $\overline{H-(K'\cap K'')} = \overline{(H-K')\cup(H-K'')} = \overline{H-K'}\cup\overline{H-K''}$, $x\in\overline{H-(K'\ K'')}$ and so $K'\cap K''\in\{K\}$. Thus an intersection of a finite number of K's is a K. Since $x\in\bar{H} = \overline{H-\emptyset}$, $\emptyset\in K$. Let $\delta\in\Delta$; let K_δ be the intersection of the K's belonging to δ; and choose $x'(\delta)\in K_\delta$. Then, if $K\in K$ and $K\in\delta$, $x'(\delta)\in K$. I say that $x'(\delta|D)$ converges to x. For if $\Delta'\subset\Delta$ is such that $x\in\overline{x'(\Delta')}$, then $H-x'(\Delta')$ is a K and no δ containing this K can belong to Δ'. Hence Δ' is not cofinal in Δ. Hence, is Δ' is cofinal in Δ, then $x\in\overline{x'(\Delta')}$. Conversely, let $x(a|\mathcal{A})$ converge to x and $x(\mathcal{A})\subset H$. Now $\mathcal{A}$ is cofinal in $\mathcal{A}$, hence (3.4) $x\in\overline{x(\mathcal{A})}\in\bar{H}$. Hence 3.7, 3.7p and 3.7pp hold.

5°. 3.10 and 3.5 imply 3.8.

If $x\in\bar{H}$, and if N is a nbd of x; then (3.5) $x\notin\overline{H-N}$; hence $H\not\subset\overline{X-N}$; hence (3.10) $H\not\subset X-N$, that is $N\cap H\neq\emptyset$. Conversely, if, for all nbds, N, of x, $H\cap N\neq\emptyset$, then, since $H\cap(X-H)=\emptyset$, $X-H$ is not a nbd of x. Hence (3.5) $x\in\overline{X-(X-H)} = \bar{H}$. So we have proved 3.8.

6°. 3.11 and 3.6 imply 3.9.

Let $x(a|\mathcal{A})$ converge to x, and let N be a nbd of x. Let $\mathcal{A}' = \{a|x(a)\notin N\}$. If $\mathcal{A}'$ were cofinal in $\mathcal{A}$, then (3.2) $x(a|\mathcal{A}')$ would converge to x, which is impossible (3.6) since $x(\mathcal{A}')\cap N = \emptyset$. Hence (II-2.4) $\mathcal{A}-\mathcal{A}'$ is residual in $\mathcal{A}$; that is, there is an a' such that $a > a'$ implies $a\notin\mathcal{A}'$, that is, $x(a)\in N$. Conversely, suppose that, for each nbd, N, of x, there is an a' such that $a > a'$ implies $x(a)\in N$. Let $\mathcal{A}'$ be cofinal in $\mathcal{A}$. If $X-x(\mathcal{A}')$ were a nbd of x; then, for some a', $a > a'$ would imply $x(a)\in X-x(\mathcal{A}')$. Since $\mathcal{A}'$ is cofinal in $\mathcal{A}$, this is impossible and $X-x(\mathcal{A}')$ is not a nbd of x. Hence (3.6) there is an $x''(b|\mathcal{B})$ such that $(X-x(\mathcal{A}'))\cap x''(\mathcal{B}) = \emptyset$, and $x''(b|\mathcal{B})$ converges to x. Hence $x''(\mathcal{B})\subset x'(\mathcal{A}')$ and (3.11), since $\mathcal{B}$ is cofinal in $\mathcal{B}$, some $x'(\delta|D)$ converges to x with $x'(\Delta)\subset x''(\mathcal{B})\subset x(\mathcal{A}')$. Since $\mathcal{A}'$ was any cofinal subsystem of $\mathcal{A}$, we have (3.11) $x(a|\mathcal{A})$ converging to x. So we have proved 3.9.

7°. 3.11 and 3.7 imply 3.14.

Let $x(a|\mathcal{A})$ converge to x, and let $\mathcal{A}'$ be cofinal in $\mathcal{A}$. Then (3.2) $x(a|\mathcal{A}')$ converges to x; hence (3.7)

$x \in \overline{x(\mathfrak{a}')}$. Conversely, suppose that for every $\mathfrak{a}'$ cofinal in $\mathfrak{a}$, $x \in \overline{x(\mathfrak{a}')}$. Then (3.7) some $x'(b|\mathfrak{B})$ converges to x , where $x'(\mathfrak{B}) \subset x(\mathfrak{a}')$. The argument proceeds precisely as in the last part of 6° , and we have $x(a|\mathfrak{a})$ converging to x . So 3.4 is proved.

8°. 3.12 and 3.8 imply 3.5.

If N is a nbd of x , then (3.8) , since $N \cap (X-N) = \emptyset$, $x \notin \overline{X-N}$. Conversely, if $x \notin \overline{X-N}$, then (3.8), for some nbd N' of x , $(X-N) \cap N' = \emptyset$. That is, $N \supset N'$; hence (3.12) N is a nbd of x . So we have proved 3.5.

9°. 3.12 and 3.9 imply 3.6, 3.6p and 3.6pp.

If N is a nbd of x , and if $x(a|\mathfrak{a})$ converges to x ; then (3.9) $x(\mathfrak{a}) \cap N \neq \emptyset$. Conversely, if H is not a nbd of x , then (3.12) H contains no nbd of x . If $\delta \in \Delta$, and $\{N\}$ is the collection of nbds of x belonging to δ , then the intersection of these N is a nbd of x , say N_δ , and is not contained in H . Choose $x'(\delta)$ so that it belongs to N_δ but not to H . Therefore, $N \in \{N\}$ implies $x'(\delta) \in N$. Now $x'(\delta|D)$ converges to x , and $H \cap x'(\Delta) = \emptyset$. So we have proved 3.6, 3.6p and 3.6pp.

These nine parts, together with 3.13 prove the various statements of the theorem.

3.15 Remarks. a) This theorem shows that at the stage of generality represented by the strong conditions the three notions of closure operator, convergence, and nbd system stand on an equal footing. From a theoretical standpoint they serve equally well as the initial concept.

b) One might remark that 3.1 and 3.10 are the same, and that this is a sort of dissymmetry. This could be easily avoided by changing some of 3.4 to 3.9, after which we could use $\overline{\emptyset} = \emptyset$, $\overline{H \cup K} \subset \overline{H} \cup \overline{K}$ in place of 3.1. However, it does not seem worth while to complicate 3.4 to 3.9 to obtain apparent symmetry. For in the best behaved (Euclidean) spaces, 3.4 to 3.9, as they stand, are the simplest relations between closure, convergence and nbds.

c) I do not say that, in practice, convergence is always as convenient and useful as closure and nbds. This is not true. But two things are true; first, there are spaces of interest to the analyst where convergence is the natural initial concept; second, convergence is a valuable tool, even in the most general space, whose neglect is wasteful in many problems.

d) The statement that 3.6 and 3.7 can be replaced by 3.6p or 3.6pp and 3.7p or 3.7pp respectively is freely expressed as

The convergence of phalanxes is sufficient to describe the topology of a space satisfying the strong conditions.

e) One result of this theorem is that, if one notion satisfies the strong condition, then we may define the other two notions so that 3.1 thru 3.12 hold. This level of generality seem a very natural one, so we make the following definition.

3.16 Definition. A topology for a set consists of a closure operator, notion of convergence and nbd system satisfying 3.1 thru 3.12. A space is a set and a topology for that set. The points of the space are the points of the set.

It is clear that either a closure operator satisfying 3.10 or a notion of convergence satisfying 3.11 or a nbd system satisfying 3.12 defines a unique topology.

A collection, $\{N_a | A\}$, of nbds of x, is a nbd basis at x, if for every nbd N of x and some a, $N_a \subset N$. A collection, $\{N_a | A\}$, of nbds of x is a nbd sub-basis at x, if for each nbd N of x and some α, $\cap\{N_a | a \in \alpha\} \subset N$. We have,

3.17 **. If $\{N_a | A\}$ is a nbd basis at x, then $\{N_b | B\}$ is a nbd basis at x if and only if, both for each a and some b, $N_b \subset N_a$, and for each b and some a, $N_a \subset N_b$.**

3.18 **. If the old nbds satisfy 3.3, and if the new nbds are derivable from (either and hence both) the convergence and the closure derivable from the old nbds, then the old nbds of x are a nbd basis at x with respect to the new nbds.**

Results similar to 3.18 dealing with closures satisfying 3.1 and convergences satisfying 3.2 can easily be stated and proved.

If a set has two topologies (which we indicate by indices 1 and 2), then topology$_1$ is finer than topology$_2$ if any (and hence all) of the following equivalent conditions hold,

3.21 $\bar{H}^{(1)} \subset \bar{H}^{(2)}$, for all H.

3.22 $x(a|\mathfrak{a})$ converges$_1$ to x implies $x(a|\mathfrak{a})$ converges$_2$ to x.

3.23 N is a nbd$_1$ of x implies N is a nbd$_2$ of x.

4. Effectiveness. We now inquire about the effectiveness of different directed systems as carriers of convergent objects. We consider convergence in spaces (3.16), where, since convergence is derivable from nbds all directed systems are equally favored.

4.1 Definition. A directed system $\mathfrak{A}$ is said to be as effective for convergence as a directed system $\mathfrak{B}$, if, whenever $x(b|\mathfrak{B})$ converges to x, there exists an $x'(a|\mathfrak{A})$ converging to x, with $x'(\mathfrak{A}) \subset x(\mathfrak{B})$.

4.2 Theorem. $\mathfrak{A}$ is as effective as $\mathfrak{B}$ if and only if $\mathfrak{A} > \mathfrak{B}$ (in the sense of II-3).

Proof. If $\mathfrak{A} > \mathfrak{B}$, there exist functions $a(b|\mathfrak{B})$ and $b(a|\mathfrak{A})$ such that $a > a(b)$ implies $b(a) > b$. If $x(b|\mathfrak{B})$ converges to x, then we set $x'(a) = x(b(a))$, thus defining $x'(a|\mathfrak{A})$. If N is a nbd of x, then (3.9), for some b', $b > b'$ implies $x(b) \in N$. But if $a > a(b')$, we have $b(a) > b'$, hence $x'(a) \in N$; hence (3.9) $x(a|\mathfrak{A})$ converges to x.

If $\mathfrak{A}$ is as effective as $\mathfrak{B}$, then we consider the following space; $X = B \cup \{\infty\}$, where $\infty \notin B$, and H is a nbd of ∞ if, for some b', $b > b'$ implies $b \in H$. The other points of X are assigned nbds in any way satisfying 3.12. The identical function $(b^*(b) = b)$ converges to ∞; hence (4.1) some $b''(a|\mathfrak{A})$ converges to ∞. Now $B(b) = \{b'|b' > b\}$ is a nbd of ∞; hence (3.9) there is an $a(b)$ such that $a > a(b)$ implies $b''(a) \in B(b)$, Hence $\mathfrak{A} > \mathfrak{B}$.

4.3 Remarks. a) This theorem provides additional support for the use of $>$ as a central relation in the theory of directed systems.

b) There is (II-5.1) a stack as effective as any particular directed system. Together with 4.2 this explains why phalanxes are sufficient for general topological purposes.

c) In the countable case (II-6) any directed system without last element is as effective as any other. This explains the usefulness of the simple sequence in the countable case.

d) If $x(n|N)$ is a simple sequence, and if we set $x'(\nu) = x(|\nu|)$, then $x'(\nu|N)$ is a phalanx which converges to those points and only those points to which $x(n|N)$ converges. The sequence and the phalanx have similar properties with regard to other notions (as "cluster point" IV-3) of a convergence nature.

If $x''(\nu|N)$ is a phalanx on a countable base (which we may as well take to be the set of the positive integers), and if we set $\nu_n = \{n'|n' \leq n\}$ and $x^*(n) = x''(\nu_n)$, then $x^*(n|N)$ converges to all the limits of $x''(\nu|N)$, but the converse need not be true.

e) The similarity of sequence and phalanx where all is countable does not (II-7) continue when we replace $\aleph_0$ by any uncountable cardinal number.

5. Relativization. If X is a space, and $Y\subset X$ then the topology of X induces a topology in Y. This occurs thru the relations,

5.1 The closure in Y, of $H\subset Y$, $\bar{H}^{(Y)}$, is equal to $\bar{H}\cap Y$.

5.2 $x(a|\mathfrak{a})$, where $x(\mathfrak{a})\subset Y$ converges in Y to $x\in Y$ if and only if $x(a|\mathfrak{a})$ converges to x in X.

5.3 The nbds of $x\in Y$ are all sets of the form $Y\cap N$ where N is a nbd of x in X.

We have,

5.4 Theorem. If X is a space (3.16), then 5.1, 5.2 and 5.3 induce a topology in Y, making Y a space.

Proof. We leave the proof to the interested reader.

Under the circumstances described in 5.4 we say that Y is a subspace of X, and that X is a superspace of Y. We shall have occasion to use,

5.5 ****. If $x\in\bar{H}$ and N is a nbd of x, then $x\in\overline{H\cap N}$.

6. Open sets and T-spaces. We begin with

6.1 ****. The following conditions on a subset U of a space X are equivalent;

6.2 $\overline{X-U}\subset X-U$.

6.3 If $x(a|\mathfrak{a})$ converges to $x\in U$, then $x(\mathfrak{a})\cap U\neq\emptyset$.

6.3p If $x(\gamma|C)$ converges to $x\in U$, then $x(\Gamma)\cap U\neq\emptyset$.

6.3pp If $x(\delta|D)$, where $D=2^X$, converges to $x\in U$, then $x(\Delta)\cap U\neq\emptyset$.

6.4 U is a nbd of all its points.

A set satisfying these equivalent conditions is open. From 6.4 we may easily prove, since every point has a nbd, that

6.5 ****. In a space X open sets satisfy;

6.6 $\emptyset$ and X are open.

6.7 The intersection of a finite number of open sets is an open set.

6.8 Any union of open sets is an open set.

We shall be really interested in open sets only when there are "enough" of them. The conditions for this are stated in

6.9 Lemma. For a space X, the following conditions are equivalent;

6.10 The open sets containing x form a nbd basis at x.

6.11 $\bar{H} \supset H$ and $\bar{\bar{H}} = \bar{H}$.

Proof. Assume 6.10. Consider a mapping of a directed system having one element onto x . Since (6.10) every nbd of x contains x , this mapping converges to x ; hence $x \in H$ implies $x \in \bar{H}$; hence $\bar{H} \supset H$. If $x \in \bar{H}$, then X-H is a nbd of x ; hence (6.10) there is an open set U , such that $x \in U \subset X-H$. Now $H \subset X-U$, so that (3.1a, 6.2) $\bar{H} \subset \overline{X-U} = \overline{H-U}$, hence $x \notin \bar{\bar{H}}$. Thus $\bar{\bar{H}} \subset \bar{H}$, and applying the first relation to $\bar{H}$, $\bar{\bar{H}} \supset \bar{H}$; hence $\bar{\bar{H}} = \bar{H}$, and 6.11 is proved.

Assume 6.11. Let N be a nbd of x , then $x \in \overline{X-N} \supset X-N$, so that $x \in N$. Now $x \notin \overline{X-N} = \overline{\overline{X-N}}$, so that $U = X-\overline{X-N}$ is a nbd of x . Now (6.2) U is open, and, since $\overline{X-N} \supset X-N$, $U \subset X-(X-N) = N$. Thus 6.10 holds.

A space satisfying these conditions is a T-space. The open sets containing x are open nbds of x . An alternate expression of 6.10 is

6.12 N is a nbd of x if and only if $x \in U \subset N$ where U is open.

We shall restrict the use of U, V, and W, to open sets, and the use of $\mathfrak{U}$, $\mathfrak{V}$, and $\mathfrak{W}$ to collections of open sets.

The reader may easily show that

6.13 **. If a family of sets satisfy 6.6, 6.7 and 6.8, and if we call them "open" and use 6.12 to define nbds, then we will obtain a T-space, in which the original family of sets is the family of open sets.**

A set, H , is closed if $\bar{H} = H$; that is (6.2), if its complement X-H is open. From 6.10, we see that

6.14 **. In a T-space $\bar{H}$ is a closed set.**

We shall often be interested in subcollections of the collection of all open set of a T-space. A subcollection is a basis (for X , for the open sets of X , etc.) if every open set is a union of sets belonging to the subcollection. A subcollection is a sub-basis (for X , etc.) if every open set is a union of sets which are themselves each intersections of a finite number of sets of the subcollection.

It is easy to see that,

6.15 **. If $\{U_a | A\}$ is a basis for X, then, for all x, $\{U_a | a \in A,\ x \in U_a\}$ is a nbd basis at x.**

6.16 **. If $\{U_a | A\}$ is a sub-basis for X, then, for all x, $\{U_a | a \in A,\ x \in U_a\}$ is a nbd sub-basis at x.**

6.17 **. If $\{N_a|A\}$ is a nbd basis at x, and $\{U_b|B\}$ is a collection of open nbds of x such that, for each a and some b, $U_b \subset N_a$; then $\{U_b|B\}$ is a nbd basis at x.**

6.18 **. If $\{U_c|C\}$ is a sub-basis for X, then the topology of X is described by the convergence of C-phalanxes. That is,**

6.19 $x \in \bar{H} \Longleftrightarrow$ **for some** $x(\gamma|C)$ **converging to x,** $x(\Gamma) \subset H$.

6.20 N is a nbd of $x \Longleftrightarrow$ **if** $x(\ |C)$ **converges to x, then** $x(\Gamma) \cap N \neq \emptyset$.

From 4.3 c), 4.3 d) and 6.18 we see that

6.21 **. The topology of a T-space with a countable sub-basis is determined by the convergence of simple sequences.**

A space is <u>discrete</u> if every subset is both open and closed. Clearly,

6.22 **. A discrete space is a T-space.**

<u>7. Continuity</u>. We have,

7.1 Theorem. If f is a function defined on the space X and taking values in the space Y, the following are equivalent,

7.2 $f(\bar{H}) \subset \overline{f(H)}$, **for all** $H \subset X$.

7.3 $x(a) \xrightarrow[\mathfrak{A}]{} x$ **implies** $f(x(a)) \xrightarrow[\mathfrak{A}]{} f(x)$.

7.3pp $x(\delta) \xrightarrow[\delta]{} x$, **where** $D = 2^X$, **implies** $f(x(\delta)) \xrightarrow[\delta]{} f(x)$.

7.4 If N is a nbd of f(x), then $f^{-1}(N)$ **is a nbd of x.**

<u>Proof</u>, 7.2 implies 7.3, for let $x(a) \xrightarrow[\mathfrak{A}]{} x$. Then (3.4), if $\mathfrak{A}'$ is cofinal in $\mathfrak{A}$, $x \in \overline{x(\mathfrak{A}')}$; hence (7.2) $f(x) \subset f\overline{(x(A'))} \subset \overline{f(xA')}$; hence (3.4) $f(x(a)) \xrightarrow[\mathfrak{A}]{} f(x)$.

7.3 obvious implies 7.3pp.

7.3pp implies 7.4, for let N be a nbd of f(x) . Then (3.6pp), whenever $y(\delta) \longrightarrow f(x)$, $N \cap y(\Delta) \neq \emptyset$. If $x(\delta) \longrightarrow x$, then (7.3pp) $f(x(\delta)) \longrightarrow f(x)$; hence $N \cap f(x(\Delta)) \neq \emptyset$, whence $f^{-1}(N) \cap x(\Delta) \neq \emptyset$. Hence (3.6pp) $f^{-1}(N)$ is a nbd of x .

7.4 implies 7.2, for let $f(x) \in f(\bar{H})$. Therefore (3.8) every nbd of x meets H . Now let N be a nbd of f(x) ; then (7.4) $f^{-1}(N)$ is a nbd of x ; hence it meets H ; hence $f(f^{-1}(N)) = N$ meets f(H) . Hence (3.8) $f(x) \in \overline{f(H)}$, and this proves $f(\bar{H}) \subset \overline{f(H)}$.

The function f is <u>continuous</u> if it satisfies these conditions. The usual properties of continuous functions can be easily derived from these conditions. Some noteworthy properties are

7.5 **. Every constant is continuous.**

7.6 **. A continuous function of a continuous function is a continuous function.**

7.7 **. The identity is a continuous function (here Y=X).**

From 6.4 we easily see that

7.8 **. If f is continuous, then**

7.9 If V is open in Y, then $f^{-1}(V)$ is open in X.
If Y is a T-space, and if 7.9 holds, then f is continuous.

If f and f^{-1} are both single-valued and continuous, then they are homeomorphisms. If f is a homeomorphism, and $f(X) = Y$; then X and Y are homeomorphic. Regarded as spaces, any two homeomorphic spaces are abstractly identical.

8. Separation Axioms. We mention two separation axioms briefly.

8.1 **. If X is a T-space, the following are equivalent;**

8.2 Every point of X is closed.

8.3 For each x, x $\bigcap\{N|N\}$ a nbd of x}.

8.4 There is a basis $\{U_a|A\}$ for X such that, if $x'\in U_a$ implies $x''\in U_a$, then x'=x".

A T-space satisfying these conditions is a T_1-space.

A space satisfying

8.5 If F' and F" are disjoint closed sets, then there are disjoint open sets U' and U", such that $F'\subset U'$ and $F''\subset U''$.

is normal. We need the following result later

8.6 **. If {U} is a basis for the normal T_1-space X, and if $x\in V$, where V is open, then there are sets U^* and U^{**} in {U} such that $x\in U^*\subset\overline{U^*}\subset U^{**}\subset\overline{U^{**}}\subset V$.**

Proof. x is closed (8.2) and X-V is closed, hence (8.5) there are disjoint U' and U" such that $x\in U'$ and $U''\subset X-V$. Now (6.15) we may choose $U^{**}\in\{U\}$, so that $x\in U^{**}\subset U'\subset X-U''$. Hence (3.1a) $\overline{U^{**}}\subset\overline{X-U''} = X-U''\subset V$, so that $x\in U^{**}\subset\overline{U^{**}}\subset V$. Applying the same argument to $x\in U^{**}$ instead of $x\in V$, we complete the proof of the lemma.

9. Historical Remarks. The theorem about effectiveness of phalanxes should be regarded as a monument to E. H. Moore, and to his faculty of grasping fundamentals. He introduced the "σ-limit" in 1915, which was essentially the convergence of phalanxes. It is true that he always restricted himself to numerically valued functions, but the basic idea was there. In 1922 Moore and Smith extracted (from the notions of sequential limit and "σ-limit") the notion of convergence of real valued functions on a directed system.

The extension to topological spaces was made by Garrett Birkhoff in 1937, who introduced the term "directed set" for a function on a directed system and who showed that convergence was adequate to topologize T_1-spaces.

It is unfortunate that the work of Moore and Smith was neglected by topologists for so many years. If it had not been neglected, the cumbersome and messy machinery of transfinite sequences and complete limit points would never have been used.

Chapter IV
COMPACTNESS

1. Introduction.
2. Special phalanxes.
3. Ultraphalanxes and topophalanxes.
4. Compactness.
5. Compact Spaces.
6. Compactification.
7. Historical Remarks.

<u>1. Introduction</u>. In this chapter we apply convergence ideas to obtain a notion of "compactness" for general spaces. This notion reduces to the "bicompactness" of Alexandroff and Urysohn for T-spaces. The important notion of an ultraphalanx is introduced.

§2 is devoted to certain set-theoretical existence theorems. In §3 we define "cluster point," "ultraphalanx" (and "topophalanx") and derive some simple properties of these concepts. In §4 we show the equivalence of many conditions and use these conditions to define "compactness." §5 contains certain results about compact spaces. In §6 we imbed spaces in compact spaces, making use of the fact that compactness may be regarded as a completeness property. We close with some historical remarks.

<u>2. Special phalanxes</u>. We begin with some set-theoretical considerations. These will lead to existence theorems for certain types of phalanxes.

A <u>covering</u> of X is a collection $\{H_a | A\}$ of subsets of X such that $\cup\{H_a | A\} = X$. A covering is a <u>partition</u> if its sets are disjoint. A covering is <u>binary</u> or <u>finite</u> if it consists of two or a finite number of sets, respectively. A subcollection $\{H_a | A'\}$, which is itself a covering is a <u>subcovering</u> of $\{H_a | A\}$.

The fundamental existence lemma is

2.1 Lemma. If $\mathfrak{B}$ is a collection of sets satisfying

2.2 $X \notin \mathfrak{B}$

2.3 $B' \subset B$ and $B \in \mathfrak{B}$ imply $B' \in \mathfrak{B}$.

2.4 $B' \in \mathfrak{B}$ and $B'' \in \mathfrak{B}$ imply $B' \cup B'' \in \mathfrak{B}$.
and if $\{\mathfrak{S}_a | A\}$ is a collection of finite coverings, then there exists a function $\varphi(a|A)$ such that, for each a , $\varphi(a)$ is a set belonging to $\mathfrak{S}_a$ and, for each α, $\bigcap\{\varphi(a) | a \in \alpha\} \notin \mathfrak{B}$

Proof. We consider the class of functions $\{\varphi\}$ each of which is; 1° defined on a subset A' of A , and such that, for $a \in A'$, $\psi(a)$ is a set of $\mathfrak{S}_a$; 2° such that $\alpha \subset A'$ implies $\bigcap\{\varphi(a)|a \in \alpha\} \notin \mathfrak{B}$. This class is not empty, for it contains (2.1) the function whose domain of definition is $\emptyset \subset A$. Since 1° and 2° are finite restrictions, there exists (I-6.4) a maximal function, which we will denote by φ , in $\{\psi\}$

Suppose that φ is not defined for $a' \in A$. Since φ is maximal, any extension of its definition to a' conflicts with 1° or 2°. Hence if $\mathfrak{S}_{a'} = (P_1, \cdots, P_m)$, where $\bigcap \{ P_i | 1 \leq i \leq m \} = X$, there must exist finite sets $\alpha_1, \cdots, \alpha_m$ such that

$$P_i \cap (\bigcap\{ \varphi(a) | a \in \alpha_i\}) \in \mathfrak{B}, \quad i = 1, \cdots, m.$$

If we set $\alpha = \bigcap\{\alpha_i \mid 1 \leq i \leq m\}$, then (2.3),

$$P_i \cap (\bigcap \varphi(a) | a \in \alpha\}) \in \mathfrak{B}, \; i = 1, \cdots, m.$$

And (2.4)

$$\bigcap\{ \varphi(a) | a \in \alpha\} = (\bigcup\{ P_i | 1 \leq i \leq m) \cap (\bigcap\{ \varphi (a) | a \in \alpha\}) \in \mathfrak{B}.$$

The last relation contradicts 2°; hence φ must have been defined for all $a \in A$, and the lemma is proved.

A function on a directed system, $x(a|\mathfrak{A})$, is <u>ultimately</u> in H if, for some a' and all $a > a'$, $x(a) \in H$. If $x(a|\mathfrak{A})$ is ultimately in H , it <u>decides for</u> H . If $x(a|\mathfrak{A})$ is ultimately in $X-H$, it <u>decides against</u> H . In either case it <u>decides about</u> H . Since $\mathfrak{A}$ is directed, we have

2.5 ****. If $x(a|\mathfrak{A})$ decides for H' and for H'' , then it decides for $H' \cap H''$.

$x(a|\mathfrak{A})$ is <u>decided about</u> a covering if it decides for at least one set of the covering. This set is unique if the covering is a partition. If the covering is not a partition, then some $x(a|\mathfrak{A})$ decides for at least two different sets of the covering. If $x(a|\mathfrak{A})$ is decided about every set of $\mathfrak{M}$, and if $\mathfrak{M}$ has a finite subcovering, then $x(a|\mathfrak{A})$ is decided about $\mathfrak{M}$. If $\mathfrak{M}$ lacks a finite subcovering, then some $x(a|\mathfrak{A})$ decides against each of its sets, and therefore is not decided about $\mathfrak{M}$.

We may rephrase III-3.9 as,

2.6 ****. $x(a|\mathfrak{A})$ converges to x if and only if it decides for every nbd of x .

We have,

2.7 Theorem. If $\mathfrak{B}$ satisfies 2.2, 2.3, and 2.4, and if $\{\mathfrak{S}_a | A\}$ is a collection of finite coverings, then there exists an A-phalanx, $x(\alpha|A)$, decided about each $\mathfrak{S}_a$. If each $\mathfrak{S}_a$ is a partition, then, if $\varphi(a)$ is the unique set of $\mathfrak{S}_a$ for which $x(\alpha|A)$ decides, we have, for each α, $\bigcap\{\varphi(a)|a\in\alpha\}\notin\mathfrak{B}$.

Proof. If $\emptyset\in\mathfrak{B}$, we consider $\mathfrak{B}^* = \mathfrak{B}\cup\{\emptyset\}$; clearly 2.2, 2.3 and 2.4 still hold. Let $\varphi(a|A)$ be the function whose existence is asserted in 2.1, and let $x(a)\in\bigcap\{\varphi(a)| a\in A\}$, this intersection is not $\emptyset$, since it does not belong to $\mathfrak{B}^*$. Then $x(\alpha|A)$ decides for $\varphi(a)$, and the theorem follows.

3. Ultraphalanxes and topophalanxes. We are now in a position to introduce some entirely new notions.

We are interested in phalanxes which have certain properties of decision, and we begin with a purely set-theoretical property. A phalanx, $x(\alpha|A)$, is an ultraphalanx (in X) if it decides about every subset of X .

We may say that an ultraphalanx is trivial if it is ultimately constant. It is clear that every ultraphalanx on a finite set is trivial. The effective (in the sense of Sierpinski) construction of a non-trivial ultraphalanx on the integers (or any other effectively enumerable set) would imply (H. E. Robbins, unpublished) the effective construction of a non-measurable (Lebesgue) set in $[0,1]$. From the present standpoint of mathematics, the ultraphalanx is a very nonconstructive tool, but a very useful one.

An important and simple property is,

3.1 Theorem. Any image of an ultraphalanx is an ultraphalanx.

Proof. Suppose that $x(\alpha|A)$ is an ultraphalanx in X , and that f is a (single-valued) function mapping X into Y . Then $f(x(\alpha)|A)$ is an ultraphalanx, for it makes the same decision about S as $x(\alpha|A)$ does about $f^{-1}(S)$.

A weaker requirement, depending on the topology of the space is also of interest. A phalanx is a topophalanx if, for every $H\subset X$, it decides about $\overline{H}$. In a T-space we may use the simpler definition; a topophalanx is a phalanx decided about each open set.

We shall apply the existence theorems of the last to these definitions. We agree that $D = 2^X$.

3.2 Theorem. If X is a space, its topology is described by the convergence of D-ultraphalanxes. That is, we have,

3.3 $x \in \bar{H}$ **if and only if some ultraphalanx** $x(\delta|D)$ **converges to** x **, with** $x(\Delta) \in H$ **.**

3.4 N **is a nbd of** x **if and only if** $N \cap x(\Delta) \neq \emptyset$ **for every ultraphalanx** $x(\delta|D)$ **converging to** x **.**

3.5 $x(a|\mathfrak{A})$ **converges to** x **if and only if, whenever** $\mathfrak{A}'$ **is cofinal in** $\mathfrak{A}$ **, there is an ultraphalanx** $x'(\delta|D)$ **converging to** x **, with** $x'(\Delta) \subset x(\mathfrak{A}')$ **.**

3.6 A function f **from** X **to** Y **is continuous if and only if, whenever** $x(\delta|D)$ **is an ultraphalanx converging to** x **,** $f(x(\delta)|D)$ **converges to** $f(x)$ **.**

Proof. We prove only 3.3; the other results follow easily from this by the methods of III-3 and III-7. To deal with 3.3, we consider the subspace (III-5) $Y = H \cup \{x\}$. If $x \in \bar{H}$, then we apply 2.7 to H , $\mathfrak{B} = \{K | K \subset H, x \notin \bar{K}\}$, and the partitions $\{\mathfrak{G}_d | D\}$, where $\mathfrak{G}_d = (H \cap L, H - H \cap L)$ and $L = d \in D = 2^X$. From III-3.10, we see that $\mathfrak{B}$ satisfies 2.2, 2.3 and 2.4. Hence, there exists a phalanx $x(\delta|D)$ which clearly is decided about every set $L = d \in D$, and which is therefore an ultraphalanx. Let N be a nbd of x in Y , then (III-3.5) $x \notin \overline{Y-N} \supset \overline{H-N}$; hence $H - N \in \mathfrak{B}$, and $x(\delta|D)$ decides for N. Hence (2.6) $x(\delta|D)$ converges to x in Y ; hence (III-5.2) it converges to x in X . Half of 3.3 is proved, the remainder follows from III-3.7.

In a similar way we may prove,

3.7 **. If** $\{U_a | A\}$ **are the open sets of the T-space** X **, then the topology of** X **is described by the convergence of A-topophalanxes.**

3.8 Remark. It is of interest to compare III-6.9 with 3.2 and 3.7. In the case of a separable metric space they assert that the topology is described by the convergence of phalanxes on a base of cardinal number $\aleph_0$, topophalanxes on a base of cardinal number $2^{\aleph_0}$, ultraphalanxes on a base of cardinal number $2^{2^{\aleph_0}}$

We now define the important notion of cluster point. x is a cluster point of $x(a|\mathfrak{A})$ if, for each $a' \in \mathfrak{A}$ and each nbd N of x , there is an $a > a'$ such that $x(a) \in N$. Intuitively this means that $x(a|\mathfrak{A})$ is not ultimately "away" from x .

3.9 **. If** $x(a|\mathfrak{A})$ **converges to** x **, then** x **is a cluster point of** $x(a|\mathfrak{A})$ **.**

3.10 **. A topophalanx or ultraphalanx converges to any cluster point(s) it possesses.**

3.11 **.** x **is a cluster point of** $x(a|\mathfrak{A})$ **if and only if** $x(a|\mathfrak{A})$ **decides against no nbd of** x **.**

3.12 Lemma. x is a cluster point of $x(\gamma|C)$ if and only if $x \in \overline{H_\gamma}$ for each γ, where $H_\gamma = \{x(\gamma')|\gamma' \supset \gamma$.

Proof. If x is a cluster point of $x(\gamma|C)$, then, since $x(\gamma|C)$ decides for H_γ, we see (3.11) that $X-H_\gamma$ is not a nbd of x. Hence (III-3.5) $x \in \overline{H_\gamma}$. Conversely, if $x \in \overline{H_\gamma}$ for all γ, and if N is a nbd of x, then $N \cap H_\gamma \neq \emptyset$ for all γ and N, which is precisely the condition that x is a cluster point of $x(\gamma|C)$.

3.13 **.** If x is a cluster point of $x(\alpha|A)$, and if $x(\alpha|A)$ is inflated, then x is a cluster point of the inflated phalanx. Conversely, a cluster point of the inflated phalanx is a cluster point of the original phalanx.

3.14 **.** If x is a cluster point of $x(a|\mathfrak{A})$ and $x(a|\mathfrak{A})$ decides for H, then $x \in \overline{H}$.

3.15 Theorem. If $x(a|A)$ is given, there exists an ultraphalanx $x'(\delta|D)$ each of whose cluster points (if any exist) is a cluster point of $x(a|\mathfrak{A})$.

Proof. Set $\mathfrak{B} = \{K|x(a|\mathfrak{A})$ decides against $K\}$; then (2.5) $\mathfrak{B}$ satisfies 2.2, 2.3 and 2.4. We apply 2.7 to $\mathfrak{B}$ and all binary partitions of X and obtain an ultraphalanx $x'(\delta|D)$. If x is a cluster point of $x'(\delta|D)$, then (3.10,2.7) $x'(\delta|D)$ decides for every nbd of x; hence none of these belong to $\mathfrak{B}$; hence (3.11) x is a cluster point of $x(a|\mathfrak{A})$.

Similarly, we see that,

3.16 **.** If $\{U_c|C\}$ are the open sets of the T-space X, and $x(a|\mathfrak{A})$ is given, there exists a topophalanx $x'(\gamma|C)$ each of whose cluster points (if any exist) is a cluster point of $x(a|\mathfrak{A})$.

4. Compactness. We begin by generalizing the notion of "subsequence." Before doing this we modify it from the classical form (both N and K are the set of positive integers), " $x(n(k)|K)$ is a subsequence of $x(n|N)$ if and only if $n(1) < n(2) < n(3) < \cdots$." to " $x(n(k)|K)$ is a subsequence of $x(n|N)$ if and only if $n(k) \geq k$." This change of definition changes none of the classical results dealing with sequences rather than with series.

We say that $x(\alpha'(\alpha)|A)$ is a subphalanx of $x(\alpha|A)$ if $\alpha'(\alpha) \supset \alpha$ for each α. It is clear that, if $x(\alpha|A)$ decides for or against H, then every subphalanx of $x(\alpha|A)$ decides likewise. Hence, from 3.11 we have,

4.0 **.** If x is a cluster point of a subphalanx of $x(\alpha|A)$, then it is a cluster point of $x(\alpha|A)$.

We are now prepared for

4.1 Theorem. The following conditions on a space X are equivalent, where $D = 2^X$

4.2 Every function $x(a|\mathfrak{A})$ on a directed system to X has a cluster point.

4.3 Every phalanx in X has a cluster point in X.

4.4 Every topophalanx in X is convergent in X.

4.5 Every D-topophalanx in X is convergent in X.

4.6 Every ultraphalanx in X is convergent in X.

4.7 Every D-ultraphalanx in X is convergent in X.

4.8 If $|A| \geq |D|$, every A-phalanx in X contains a subphalanx converging in X.

4.9 Every phalanx in X can be inflated to a phalanx containing a subphalanx converging in X.

4.10 If $\{H_c | C\}$ has the property that, for all $\gamma \subset C$, $\cap\{H_c | c \in \gamma\} \neq \emptyset$. then $\cap\{\bar{H}_c | c \in C\} \neq \emptyset$.

Proof. In view of 3.9 and 3.10, 4.2 implies 4.3, 4.4, 4.5, 4.6, and 4.7. Clearly, 4.2 or 4.3 or 4.4 or 4.5 or 4.6 implies 4.7. Thus to prove all these conditions equivalent we need only show that 4.7 implies 4.2, which follows immediately from 3.15 and 3.10.

We now show that 4.3 and 4.10 are equivalent. Assume 4.3, and take $x(\gamma) \in \cap\{H_c | c \in \gamma\}$; then (4.3) $x(\gamma | C)$ must have a cluster point x. Now, for $c \in \gamma$, $x(\gamma) \in H_c$; hence (3.12) $x \in \bar{H}_c$ and we have proved 4.10. Assume 4.10, and let $x(\gamma | C)$ be any phalanx. Let $B = \Gamma$, and let $H_b = H_\gamma = \{x(\gamma') | \gamma' \supset \gamma\}$, where $b = \gamma$. If $\gamma(\beta) = \cup\{\gamma | \gamma = b \in \beta\}$, then $x(\gamma(\beta)) \in \cap\{H_b | b \in \beta\}$. Hence, for all β, $\cap\{H_b | b \in \beta\} \neq \emptyset$; hence (4.10) $\cap\{\bar{H}_b | b \in B\} \neq \emptyset$. If $x \in \cap\{\bar{H}_b | B\} = \cap\{\bar{H}_\gamma | \gamma \in \Gamma\}$, then (3.12) x is a cluster point of $x(\beta | B)$.

It remains to be shown that 4.3, 4.8 and 4.9 are equivalent. Clearly 4.8 implies 4.9; and (3.13, 3.9, 4.0) we see that 4.9 implies 4.3. Assume 4.3, and let $x(\alpha | A)$ be an A-phalanx, where $|A| \geq |D|$; then (4.3) $x(\alpha | A)$ has a cluster point x. Let $B \subset D$ consist of the nbds of x; then, since $|B| \leq |D| \leq |A|$, there is a 1-1 correspondance $a = f(b)$ between B and a subset of A. For each α, $f^{-1}(\alpha)$ is finite, and we may choose $\alpha'(\alpha) \supset \alpha$ so that $x(\alpha'(\alpha)) \in \cap\{N | N = a \in f^{-1}(\alpha)\}$ (for this intersection is a nbd of a cluster point of $x(\alpha | A)$). Then $x(\alpha'(\alpha) | A)$ is a subphalanx of $x(\alpha | A)$, and this subphalanx obviously converges to x. Thus 4.8 is proved, and the proof of the equivalence is complete.

4.11 Definition. We call a space satisfying 4.2 thru 4.10 compact.

By the same methods we may prove,

4.12 ****. In a T-space, 4.2 thru 4.10 are equivalent to

4.13 Each open (that is, made up of open sets) covering of X possesses a finite subcovering.

4.14 If $\{U_a | A\}$ are the open sets of X, then every A-topophalanx in X is convergent in X.

4.15 If $\{U_c | C\}$ is a sub-basis for X, then every B-phalanx, where $|B| \geq |C|$, has a convergent subphalanx.

4.16 Remarks. a) The important conditions are 4.3, 4.6, 4.10 and in T-spaces 4.13. Condition 4.8 contains (see b) below) as a special case a well-known result, but 4.6 is more useful in applications.

b) Suppose that we have a T-space with a countable basis, and that B is a countable set. Then we know that the space is compact if and only if every B-phalanx has a convergent subphalanx. Proceeding as in III-4.3, we easily see that this is equivalent to the classical condition that every sequence should have a convergent subsequence.

c) We stress the fact that 4.6 exhibits compactness as a completeness property. It is reasonable to say that compactness is the ultimate completeness property of a topological nature.

d) We note again that what we call "compactness" reduces in T-spaces to what Alexandroff and Urysohn called "bicompactness."

From 4.6 and III-3.22 we easily see that,

4.17 ****. If X is a space in topology_1 and in topology_2, if topology_1 is finer than topology_2, and if X is compact in topology_1; then X is compact in topology_2.

5. Compact Spaces. A space Y is a continuous image of a space X if there is a continuous function f such that $f(X) = Y$. From 4.6, 3.1 and 3.6 we have

5.1 ****. A continuous image of a compact space is compact.

5.2 Remark. This result is distinct from the classical one in which "compact" or "bicompact" has a different meaning. The content of the classical theorem follows from this theorem and the fact that a continuous image of a T-space is a T-space.

From 4.3, 3.14, 3.3 and 4.6 we easily see that

5.3 **. A subspace Y of a compact space X is compact if and only if Y is a closed subset of X .**

We conclude this section with an interesting description of the topology of a compact space. Let us begin with a set X. Let Q consist of all the 2^X-ultraphalanxes in X . Let P consist of the equivalence classes generated in Q by the relation $\sim$. Here $q_0 \sim q'$ if there is a chain $q_0, q_1, \cdots,$ $q_n = q'$ such that, for each k , either q_{k-1} is a subphalanx of q_k or q_k is a subphalanx of q_{k-1} .

If X were a space, then (III-3.2, 3.9) if one $q \in p'$ converges to x then all $q' \in p'$ also converge to x . So we may define $f(p) \in 2^X$ as the set of x to which any (and hence all) $q \in p$ converge. It is clear (3.2) that $f(p|P)$ describes the topology of X .

We state the following without proof.

5.4 If X is finite, then every function from P to 2^X determines the topology of a space whose points are the point of X .

5.5 If X is neither finite or countably infinite, then <u>not every</u> function from P to 2^X determines a space.

5.6 If f on P to 2^X determines a space, then the space is compact if and only if $\emptyset$ is not a value of f .

Some readers may be interested in proving these statements and in proving that starting from a set containing a finite number, n , of distinguishable points we may construct the following numbers of spaces with the following properties.

spaces,	2^{n^2} ,
compact spaces,	2^{n^2-n} ,
spaces in which limits are unique,	$(n+1)^n$,
compact spaces in which limits are unique,	n^n ,
T_1-spaces,	1 (this space is normal and compact).

<u>6. Compactification</u>. We have pointed out (4.16c) that compactness is a completeness property. It is natural, therefore, to try to compactify spaces by "completing" them. It is true that we may do this trivially, for example, let us prove

6.1 Lemma. Every space X may be imbedded in a compact space.

Proof. Let $Y = X \cup \{y'\}$, and let Y be the only nbd of y' . It is easy to see (3.12, III-3.16) that Y is a space. Then every ultraphalanx in Y converges to y' , and (4.6) Y is therefore compact.

However, less trivial completion procedures lead to more interesting completions with special properties.

Suppose that X is a space. Let Z consist of all D-ultraphalanxes in X . Let $Y = X \cup Z$. We define a mapping, $H \to H^*$, of 2^X on 2^Y by

6.2 $H^* = \{x \mid x \in H\} \cup \{z \mid z$ is ultimately in $H\}$.

Clearly,

6.3 $H^* \cap X = H$,

and

6.4 $\emptyset^* = \emptyset$, and $H \subset K$ implies $H^* \subset K^*$.

Now (2.5) we see that

6.5 $(H \cap K)^* = H^* \cap K^*$.

Since every ultraphalanx is decided about the binary partition $(H, X-H)$, we have

6.6 $(X-H)^* = Y-H^* = X^*-H^*$.

Combining 6.5 and 6.6, we have

6.7 $(H \cup K)^* = H^* \cup K^*$,

and

6.8 $(H-K)^* = H^*-K^*$.

Now we define the nbds in Y . The nbds of x are all sets $S \subset Y$, with $x \in N \subset S$, where N is a nbd of x in X . The nbds of z are all sets $S \subset Y$, where, for some H , $z \in H^* \subset S$.

It will be easy for the reader to show that III-3.12 and III-5.3 hold, so that Y is a subspace of X . We now prove,

6.9 Lemma. Y is compact.

Proof. Let $y(\gamma|C)$ be an ultraphalanx in Y , and define $\{T_d | D\}$, where $D = 2^X$, by

6.10 $T_d = \begin{cases} H, & \text{if } y(\gamma|C) \text{ decides for } H^* , \\ X-H, & \text{if } y(\gamma|C) \text{ decides against } H^* . \end{cases}$

Now $y(\gamma|C)$ decides for T_d^* and hence (6.5, 2.5) for $\cap\{T_d^*|d\in\delta\} = (\cap\{T_d|d\in\delta\})^*$. Hence (6.4) $\cap\{T_d|d\in\delta\} \neq \emptyset$, and we may choose $x(\delta)\in\cap\{T_d|\delta\}$. Now $z' = x(\delta\in D)\in Z$, and $y(\gamma|C)$ converges to z'. For let S be a nbd of z', then, for some H, $S\supset H^*\ni z'$. Let $d = H$; then $T_d = H$, for otherwise $z'\in(X-H)^* = Y-H^*$. Hence $y(\gamma|C)$ decides for $H^*\subset S$, and (4.6) the lemma is proved.

We now have,

6.11 Theorem. **Any space X is an open subset of a compact space Y.**

Proof. Y is compact (6.9) and we see that X is a nbd in Y of each of its points; hence (III-6.4) the theorem follows.

In a certain sense, Y is a universal compactification of X, for we have,

6.12 Theorem. **If W is a superspace of X in which limits are unique (thus, limits in X must be unique), if $W = \bar{X}$, and if W is compact, then W is a continuous image of Y by a function which is the identity on X.**

Proof. We define f as follows; $f(x) = x$; if $z = x(\delta|D)$, then $f(z)$ is the unique limit of $x(\delta|D)$ in W. $f(Y)$ covers W, since each $w\in W-X$ is the limit in W of an ultraphalanx $x(\delta|D)$.

Suppose that $y(\gamma|C)$ is an ultraphalanx converging to y. If $y\in X$, then, since X is a nbd of Y, $y(\gamma|C)$ is ultimately in X; hence $f(y) = y$ and ultimately $f(y(\gamma))$ $= y(\gamma)$. It is clear that, in this case, $f(y(\gamma)|C)$ converges to $f(y)$. If $y\in Z$, then define $\{T_d|D\}$ and $x(\delta|D)$ as in the proof of 6.9. Let N be a nbd in Y of $y\in Z$; then, for some H, $y\in H^*\subset N$. Since $y(\gamma|C)$ converges to y, it is ultimately in H^*. Hence $x(\delta|D)$ is ultimately in $H\subset H^*$, so that $x(\delta|D)$ converges to y in the space Y and hence to $f(y)$ in the space W. Let N be a nbd of $f(y)$ in W. Then $x(\delta|D)$ is ultimately in N, and so is $y(\gamma|C)$ which is ultimately equal to $f(y(\gamma)|C)$. Hence $f(y(\gamma)|C)$ converges to $f(y)$. Hence (3.6) f is continuous and the theorem is proved.

This method of embedding X in a compact space is not only interesting, but is useful in proving

6.13 Theorem. **If X is a T-space, then there exists a compact T-space $Y\supset X$ and a one to one mapping of the open sets of X on a basis for the open sets of Y which preserves inclusion, finite intersection and finite union, and whose inverse is the natural mapping from a space to a subspace.**

Proof. Let X be a T-space, we proceed as above, but we introduce a different topology in Y. If $\{U_a | A\}$ are the open sets of X, then we define $\{U_a^* | A\}$ as a basis for the open sets of Y. It is easy to see (III-3.23) that the old topology of Y is finer than the present topology of Y; hence (4.17, 6.9) Y is compact in the present topology. The remainder of the theorem follows from 6.3 thru 6.8.

We may modify Y in another way; we define $\sim$ on Z by $z' \sim z''$ if $z' \in H^*$ implies $z'' \in H^*$. We see (6.6) that this is an equivalence relation. Let R be the set of equivalence classes. Let $P = X \cup R$. Then, if we take nbds in P in the way we first did in Y, we see that P is a continuous image of Y, and hence compact. If we examine R, we see that $\{H^* \cap R | H \subset X\}$ is a basis for its open sets and that it is a T_1-space. Thus we have

6.14 **. If X is a space, then it is open in a compact superspace P, where $R = P - X$ is a T_1space.**

A simple and intuitively interesting corollary of 6.1 is,

6.15 Theorem. A space X is compact if and only if

6.16 Whenever Y is a superspace of X, and a phalanx $x(\alpha | A)$ converges to a point in Y, then it has a cluster point in X.

Proof. The condition is necessary by 4.3 Since (6.1) X can be embedded in a space in which every ultraphalanx is convergent, 6.16 implies that every ultraphalanx in X has a cluster point in X; hence converges in X; hence (4.3) X is compact.

7. Historical Remarks. The term "compact" was introduced by Fréchet in his important paper of 1906. The motivation of the definition is explained in his book of 1928 (p. 69) as follows. "Il faut encore compléter cet énoncé primitif pour obtenir une propriété caractéristique des ensembles linéaires bornés. . . . . On voit immédiatement qu'il est naturel d'essayer de tirer de cet énoncé la definition cherchée des ensembles compacts." These remarks apply equally well to a) Fréchet's original definition (every infinite set has a nonvoid derivative), b) Alexandroff and Urysohn's definition of "bicompactness" (Our 4.13 and other equivalent forms), c) our definition of "compactness."

We note Fréchet's remarks (1921, p. 350) on another change of definition. "Ces changements peuvent amener une confusion momentanée, mais correspondent à un progrès de la théorie."

The term "bicompact" was introduced by Alexandroff and Urysohn in 1923 for three equivalent properties of a T-space; our 4.3, a special form of 4.10, and a condition concerning transfinite sequences and complete limit points. This notion has been of great service to topology. We recall that "compact" in Fréchet's sense and "bicompact" in the sense of Alexandroff and Urysohn are equivalent for metrizable spaces.

In the last few years, the younger French school of mathematicians have been using "compact" in the sense of Alexandroff and Urysohn's "bicompact."

I have chosen the term "compact" for the general notion set forth in §4, since this seems to be the important notion. I suggest the following additional terminology. "sequentially compact" = every sequence contains a convergent subsequence. "countably compact" = Fréchet's "compact" = every infinite set has a non-empty derivative (the term "countably" is justified by the fact that we may clearly restrict ourselves to countably infinite sets).

Chapter V
NORMALITY

1. Introduction.
2. Calculus of Coverings.
3. Sequences of Coverings.
4. Normal Spaces.
5. Star-finiteness.
6. Typical Refinement.
7. Special pseudo-écarts and pseudo-metrics.
8. Metrization.
9. Continuous Real Valued Functions.
10. Historical Remarks.

1. Introduction. In this chapter we develop relations between pairs and sequences of open coverings of a space. We then apply these relations to the existence of pseudo-metrics, metrics, and certain other types of continuous real-valued functions (Generalizations of Urysohn's Lemma).

In §2 we develop the formal calculus of coverings. In §3 we define normal sequences of coverings and normal coverings. In §4 we connect the notions of normal covering and normal space. In §§5-6 we obtain some results about star-finite coverings. In §7 we discuss pseudo-metrics, in §8 metrics, and in §9 continuous real-valued functions. In §10 we close with some historical remarks.

2. Calculus of Coverings. We shall develop some of the more formal properties of coverings. We recall that a collection, $\mathfrak{M} = \{M_a | A\}$ of subsets of X is a covering (of X) if $X = \cup \{M_a | A\}$. A covering $\mathfrak{M}$ is a refinement of a covering $\mathfrak{L}$, written $\mathfrak{M} < \mathfrak{L}$, if each set M of $\mathfrak{M}$ is contained in some set L of $\mathfrak{L}$. This is clearly a transitive relation ordering the class of all coverings.

We introduce two important operations on classes of coverings, $\wedge$ and $\vee$. $\wedge\{\mathfrak{M}_a | A\}$, sometimes abbreviated $\wedge \mathfrak{M}_a$ or $\wedge_a \mathfrak{M}_a$, consists of all sets $\cap \{M_a | A\}$, where $M_a \in \mathfrak{M}_a$. $\vee\{\mathfrak{M}_a | A\}$, sometimes abbreviated $\vee \mathfrak{M}_a$ or $\vee_a \mathfrak{M}_a$, consists of all sets M, where $M \in \mathfrak{M}_a$ for some a. Clearly $\vee$ and $\wedge$ preserve the property of being a covering. We (somewhat improperly) call $\wedge$ and $\vee$ intersection and union respectively.

We clearly have,

2.1 **.** $\mathfrak{M}_a < V\{\mathfrak{M}_a | A\}$; **in particular** $\mathfrak{M} < \mathfrak{L}\mathfrak{M}$.

2.2 **.** $\mathfrak{M}_a < \mathfrak{L}$ **for all** $a \in A$ **implies** $V\{\mathfrak{M}_a | A\} < \mathfrak{L}$.

2.3 **.** $\Lambda\{\mathfrak{M}_a | A\} < \mathfrak{M}_a$; **in particular** $\mathfrak{L} \wedge \mathfrak{M} < \mathfrak{M}$.

2.4 **.** $\mathfrak{L} < \mathfrak{M}_a$ **for all** $a \in A$ **implies** $\mathfrak{L} < \Lambda\{\mathfrak{M}_a | A\}$.

We observe that these relations imply that the coverings of X and the relation of refinement make up a complete trellis in the sense of I-3.

If $\mathfrak{M} < \mathfrak{L} < \mathfrak{M}$, we write $\mathfrak{L} \sim \mathfrak{M}$ and say that $\mathfrak{L}$ and $\mathfrak{M}$ are <u>equivalent</u>.

<u>Associated</u> with $\mathfrak{M} = \{M_a | A\}$ there are the binary coverings $\{\mathfrak{M}_{(a)} | A\}$, where $\mathfrak{M}_{(a)}$ consists of the two sets M_a and $\cup\{M_{a'} | a' \neq a\}$. We see that

2.5 **.** $\mathfrak{M}$ **is equivalent to** $\Lambda\{\mathfrak{M}_{(a)} | A\}$, **the intersection of its associated binary coverings.**

We now introduce the notion of the <u>star</u> $S(H,\mathfrak{M})$ of a set H in a covering $\mathfrak{M}$, which is the union of all the sets of $\mathfrak{M}$ meeting H. Clearly,

2.6 **.** $H \subset K$ **implies** $S(H,\mathfrak{M}) \subset S(K,\mathfrak{M})$.

2.7 **.** $\mathfrak{M} < \mathfrak{N}$ **implies** $S(H,\mathfrak{M}) \subset S(H,\mathfrak{N})$.

2.8 **.** $S(H,\Lambda\{\mathfrak{M}_a | A\}) = \cap\{S(H,\mathfrak{M}_a) | A\}$.

2.9 **.** $S(H,V\{\mathfrak{M}_a | A\}) = \cup\{S(H,\mathfrak{M}_a) | A\}$.

We write $S(x,\mathfrak{M})$ for $S(\{x\},\mathfrak{M})$ and we constantly make use of the equivalence of

a) x and y are in a single set of $\mathfrak{M}$,
b) $x \in S(y,\mathfrak{M})$,
c) $y \in S(x,\mathfrak{M})$.

We define <u>iterated stars</u> by

2.10 $S^n(H,\mathfrak{M}) = S(S^{n-1}(H,\mathfrak{M}),\mathfrak{M})$.

We now define two important operations on coverings.
$\mathfrak{M}^*$ consists of the sets $S(M,\mathfrak{M})$, where $M \in \mathfrak{M}$.
$\mathfrak{M}^\Delta$ consists of the sets $S(x,\mathfrak{M})$, where $x \in X$. We see easily that we have

2.11 **.** $\mathfrak{M} < \mathfrak{N}$ **implies both** $\mathfrak{M}^* < \mathfrak{N}^*$ **and** $\mathfrak{M}^\Delta < \mathfrak{N}^\Delta$.

2.12 **.** $\mathfrak{M}_a^* < \mathfrak{N}_a$ **for each** $a \in A$ **implies** $(\Lambda\{\mathfrak{M}_a | A\})^* < \Lambda\{\mathfrak{N}_a | A\}$.

2.13 **.** $S(S(H,\mathfrak{M}),\mathfrak{M})=S^2(H,\mathfrak{M})=S(H,\mathfrak{M}^\Delta)$.

2.14 **.** $\mathfrak{M}^{\Delta\Delta}=\{S^2(x,\mathfrak{M})\mid x\in X$.

2.15 **.** $\mathfrak{M}^{\Delta\ldots\Delta\,(n+1\text{ times})}=\{S^{2^n}(x,\mathfrak{M})\mid x\in X\}$.

We now have the important

2.16 Lemma. $\mathfrak{M}^\Delta<\mathfrak{M}^*<\mathfrak{M}^{\Delta\Delta}$.

Proof. Each x' belongs to some $M'\in\mathfrak{M}$: hence (2.6) $S(x',\mathfrak{M})\subset S(M',\mathfrak{M})$. We have shown that $\mathfrak{M}^\Delta<\mathfrak{M}^*$.

Let $M^*\in\mathfrak{M}^*$; then $M^*=S(M',\mathfrak{M})$ for some $M'\in\mathfrak{M}$. Let $x'\in M'$; then $M'\subset S(x',\mathfrak{M})$. Now (2.6, 2.7) $M^*=S(M',\mathfrak{M})\subset S(S(x',\mathfrak{M}),\mathfrak{M})=S^2(x',\mathfrak{M})$, which belongs to $\mathfrak{M}^{\Delta\Delta}$ Hence $\mathfrak{M}^*<\mathfrak{M}^{\Delta\Delta}$

We define the important relations $*\!<$ and $\triangleleft$ by $\mathfrak{M}*\!<\mathfrak{L}$ if $\mathfrak{M}^*<\mathfrak{L}$, and $\mathfrak{M}\triangleleft\mathfrak{L}$ if $\mathfrak{M}^\Delta<\mathfrak{L}$. Where no confusion can arise we write $\mathfrak{M}*\!<\mathfrak{L}$. If $\mathfrak{M}*\!<\mathfrak{L}$, then $\mathfrak{M}$ is a <u>star-refinement</u> of $\mathfrak{L}$. If $\mathfrak{M}\triangleleft\mathfrak{L}$, then $\mathfrak{M}$ is a <u>$\triangleleft$-refinement</u> of $\mathfrak{L}$.

The <u>restriction</u> $H\circ\mathfrak{M}$ of $\mathfrak{M}$ to H is a covering of X consisting of the set $S(X-H,\mathfrak{M})$, and those sets of $\mathfrak{M}$ included in H . We clearly have

2.17 **.** $\mathfrak{M}<H\circ\mathfrak{M}<(H,S(X-H,\mathfrak{M}))$.

2.18 **.** If $\mathfrak{M}<\mathfrak{L}$ and $H\supset K$, then $H\circ\mathfrak{M}<K\circ\mathfrak{L}$.

2.19 Lemma. If $\mathfrak{M}^*<\mathfrak{L}$, then $H\circ\mathfrak{M}^*<H\circ\mathfrak{L}$.

Proof. Let $M'\in H\circ\mathfrak{M}$. If $M'=S(X-H,\mathfrak{M})$, then (2.13, 2.16) $S(M',H\circ\mathfrak{M})=S(S(X-H,\mathfrak{M}),\mathfrak{M})\subset S(X-H,\mathfrak{L})$, which belongs to $H\circ\mathfrak{L}$. If $M'\in\mathfrak{M}$ and meets $S(X-H,\mathfrak{M})$, then $S(M',H\circ\mathfrak{M})$ $=S(M',\mathfrak{M})\cup S(X-H,\mathfrak{M})$. Further, we see that $S(M',\mathfrak{M})$, which $\subset L'\in\mathfrak{L}$, meets $X-H$; so that $S(M',\mathfrak{M})\cup S(X-H,\mathfrak{M})\subset S(X-H,\mathfrak{L})$, which belongs to $H\circ\mathfrak{L}$. Finally, if $M'\in\mathfrak{M}$ and does not meet $S(X-H,\mathfrak{M})$, then $S(M',H\circ\mathfrak{M})=S(M',\mathfrak{M})$ which surely is contained in a set of $\mathfrak{L}$, and hence of $H\circ\mathfrak{L}$. Thus the lemma is proved.

Another sort of operation which must be carefully distinguished from the last is that of <u>intersecting</u> with H . The result $H\cap\mathfrak{M}$ is the covering of H (not of X) made up of the sets $H\cap M$, where $M\in\mathfrak{M}$. Clearly,

2,20 **.** $\mathfrak{M}<\mathfrak{L}$ implies $H\cap\mathfrak{M}<H\cap\mathfrak{L}$.

2.21 **.** $\mathfrak{M}*\!<\mathfrak{L}$ implies $H\cap\mathfrak{M}^*<H\cap\mathfrak{L}$.

A covering is finite if it consists of a finite number of sets. A covering is star-finite if each of its sets meets a finite (but not necessarily bounded) number of its other sets.

2.22 **. If $\mathfrak{M}$ is star-finite, then so are $\mathfrak{M}^*$, $\mathfrak{M}^{**}$, $\cdots$.**

A collection of coverings is star-finite if the union of each pair of these covering is star-finite.

2.23 **. The union of any finite number of coverings of a star-finite collection of coverings is star-finite.**

2.24 **. A sequence $\{\mathfrak{M}_n\}$ of coverings is star-finite if and only if $\mathfrak{M}_{n+1} \vee \mathfrak{M}_n$ is star-finite for each n.**

We extend the notion of equivalence to classes of coverings. $\{\mathfrak{M}_a | A\}$ is equivalent to $\{\mathfrak{L}_b | B\}$ if both for each a and some b $\mathfrak{L}_b < \mathfrak{M}_a$, and for each b and some a $\mathfrak{M}_a < \mathfrak{L}_b$.

We shall have occasion to consider the inverse images of coverings. We easily see that

2.25 **. If f is single valued, then $\mathfrak{M} < \mathfrak{L}$ implies $f^{-1}(\mathfrak{M})^* < f^{-1}(\mathfrak{L})$.**

2.26 **. If f is single valued, then $\mathfrak{M} \overset{*}{<} \mathfrak{L}$ implies $f^{-1}(\mathfrak{M}) \overset{*}{<} f^{-1}(\mathfrak{L})$.**

3. Sequences of Coverings. For the remainder of this chapter and in the succeeding chapters we consider only open coverings (that is, coverings made up of open sets). These we will refer to merely as "coverings." The reader may easily verify that $\wedge$ (of a finite number of coverings), $\vee$, $*$, $\triangleleft$, H^o, and H^n are operations taking open coverings into open coverings.

A sequence $\{\mathfrak{U}_n\}$ is a normal sequence, if $\mathfrak{U}_{n+1} \overset{*}{<} \mathfrak{U}_n$ for each n . A covering $\mathfrak{U}$ is normal if a normal sequence $\{\mathfrak{U}_n\}$ exists with $\mathfrak{U}_1 < \mathfrak{U}$.

3.1 Lemma. Every covering in a normal sequence is normal. The intersection of a finite number of normal coverings is normal.

Proof. Let $\{\mathfrak{V}_n\}$ be a normal sequence and consider $\mathfrak{V}_k$. Let $\mathfrak{U}_n = \mathfrak{V}_{n+k}$; then $\{\mathfrak{U}_n\}$ is a normal sequence, and $\mathfrak{U}_1 = \mathfrak{V}_{k+1} < \mathfrak{V}_k$; hence $\mathfrak{V}_k$ is normal.

Let $\{\mathfrak{U}_a\}$ be a finite collection of normal coverings. $\mathfrak{U}_a$ is normal, so there is a normal sequence $\{\mathfrak{U}_{a,n}\}$ with $\mathfrak{U}_{a,1} < \mathfrak{U}_a$. Now (2.12, 2.3, 2.4), if $\mathfrak{V}_n = \wedge \{ \mathfrak{U}_{a,n} | a \in A \}$, then $\mathfrak{V}_{n+1} \overset{*}{<} \mathfrak{V}_n$ and $\mathfrak{V}_1 < \wedge \{ \mathfrak{U}_a | A \}$. Hence $\{\mathfrak{V}_n\}$ is a normal sequence, and $\wedge \{\mathfrak{U}_a | A\}$ is normal.

We easily see (2.16) that

3.2 **. If $\mathfrak{U}_{n+1} \lhd \mathfrak{U}_n$, then $\{\mathfrak{U}_{2^n}\}$ is a normal sequence.**

3.3 Lemma. If $\{\mathfrak{U}_n\}$ satisfy

3.4 $x \in S(y, \mathfrak{U}^{\Delta}_{n+1})$ and $z \in S(y, \mathfrak{U}^{\Delta}_{n+1})$ imply $x \in S(z, \mathfrak{U}^{\Delta}_n)$,

then

3.5 $S(x, \mathfrak{U}^{\Delta}_{n+1}) < S(x, \mathfrak{U}_n)$,

and each $\mathfrak{U}^{\Delta}_n$ is a normal covering.

Proof. Let $z \in S(x, \mathfrak{U}^{\Delta}_{n+1})$; then, for some y, $S(y, \mathfrak{U}_{n+1}) \in \mathfrak{U}^{\Delta}_{n+1}$ contains both x and z. Hence (3.4) $x \in S(z, \mathfrak{U}_n)$ and 3.5 is proved. Since $S(x, \mathfrak{U}_n) \in \mathfrak{U}^{\Delta}_n$ we see that $\mathfrak{U}^{\Delta}_{n+1} \lhd \mathfrak{U}^{\Delta}_n$. Hence (3.2) $\{\mathfrak{U}^{\Delta}_{r+2n} | N\}$ is a normal sequence, and since $\mathfrak{U}^{\Delta}_{r+2} < \mathfrak{U}^{\Delta}_r$, $\mathfrak{U}^{\Delta}_r$ is normal.

From 2.25, 2.26, and III-7.8 we easily see that

3.6 **. If f is continuous on X to Y, and if $\{\mathfrak{U}_n\}$ is a normal sequence in Y, then $\{f^{-1}(\mathfrak{U}_n)\}$ is a normal sequence in X.**

3.7 **. If f is continuous on X to Y and if $\mathfrak{U}$ is normal in Y, then $f^{-1}(\mathfrak{U})$ is normal in X.**

4. Normal Spaces. We have

4.1 Theorem. The following conditions on a space X are equivalent:

4.2 Every binary (open) covering of X has a (finite and open) $\lhd$-refinement.

4.3 Every finite (open) covering of X has a (finite and open) $\lhd$-refinement.

4.4 Every finite (open) covering of X is normal.

4.5 X is a normal space.

Proof. Any finite open covering is equivalent to a finite intersection of binary coverings (2.5), each of these binary coverings has a finite $\lhd$-refinement (4.2); the intersection of these finite coverings is a $\lhd$-refinement of the given finite covering (2.11). Thus 4.2 implies 4.3.

Let $\mathfrak{B}_1$ be a finite covering of X ; by successive applications of 4.3 we obtain finite coverings, $\mathfrak{B}_r$, $n = 2, 3, \cdots$, such that $\mathfrak{B}_{n+1} \lhd \mathfrak{B}_n$; hence (3.2) $\{\mathfrak{B}_{2n}\}$ is a normal sequence and, since $\mathfrak{B}_2 < \mathfrak{B}_1$, $\mathfrak{B}_1$ is normal. Thus 4.3 implies 4.4.

Clearly 4.4 implies 4.2; hence 4.2, 4.3, and 4.4 are equivalent.

If X is a normal space (III-8.5), and if $\{U',U''\}$ is a covering of X, then $X-U'$ and $X-U''$ are disjoint and closed, so that V' and V'' exist with $X-U' \subset V'$, $X-U'' \subset V''$, $V' \cap V'' = \emptyset$. Then $\{V',V'',U' \cap U''\}$ is a finite covering, which is a $\triangleleft$-refinement of $\{U',U''\}$. Thus 4.2 holds.

If 4.2 holds, and if H and K are disjoint closed sets, then $\{X-H,X-K\}$ is a binary open covering. If $\mathfrak{U}$ is a $\triangleleft$-refinement of $\{X-H,X-K\}$, then for no x does $S(x,\mathfrak{U})$ meet both H and K, so that $S(H,\mathfrak{U})$ and $S(K,\mathfrak{U})$ are disjoint open sets, containing H and K respectively. Thus X is normal.

Hence 4.2 and 4.5 are equivalent and the theorem is proved.

4.6 Lemma. If F' and F'' are disjoint closed sets in a normal space, then there exists a finite normal covering $\mathfrak{B}$ such that F' and $S(F'',\mathfrak{B})$ are disjoint sets.

Proof. If $\mathfrak{B} \triangleleft \{X-F',X-F''\}$ the reader may show that $F' \cap S(F'',\mathfrak{B}) = \emptyset$.

If $\{U\}$ is a particular basis for X, then a covering $\{U',X-U''\}$ is a basic binary covering of X if U' and U'' belong to $\{U\}$.

4.7 Lemma. If X is a normal T_1-space, and $\{U\}$ is a given basis for X, then, for all x, $\{S(x,\mathfrak{U}) | \mathfrak{U}$ a basic binary covering$\}$ is a nbd basis at x.

If the given basis is countable, then there are a countable number of basic binary coverings.

Proof. If $x \in U$, then (III-8.6) there exist sets U' and U'' of the given basis, with $x \in U'' \subset \overline{U''} \subset U' \subset \overline{U'} \subset U$. Now $\mathfrak{U} = \{U',X-\overline{U''}\}$ is a basic binary covering, and $S(x,\mathfrak{U}) = U' \subset U$.

There are a countable set of pairs (U',U'') of basic open sets; a fortiori there are a countable set of basic binary coverings.

5. Star-finiteness. If A is infinite, then $\wedge\{\mathfrak{U}_a | A\}$ need not be an open covering. However, we have,

5.1 Lemma. If $\{V_a | A\}$ is a star-finite open covering, and if each $\mathfrak{U}_a$ is open, then $\mathfrak{R}=\{\mathfrak{R}_a | A\}$, where $\mathfrak{R}_a = V_a \cdot \mathfrak{U}_a$ is an open covering.

Proof. Let $x \in R = \wedge_a R_a$, where $R_a \in \mathfrak{R}$ and each $R_a \in \mathfrak{R}$. Then, for some a', $x \in V_{a'}$; let α contain all a for which $V_a \cap V_{a'} \neq \emptyset$. If $a \notin \alpha$, the only set of $\mathfrak{R}_a = V_a \circ \mathfrak{U}_a$ meeting $V_{a'}$ is $S(X-V_a, \mathfrak{U}_a)$; hence, for $a \notin \alpha$, $R_a = S(X-V_a, \mathfrak{U}_a) \supset V_{a'}$. Now (III-6.7) $W = V_{a'} \cap (\cap \{R_a | a \in \alpha)$ is open, and $x \in W \subset R$. Hence R is a nbd of each of its points, and (III-6.4) open. This proves the lemma.

5.2 Lemma. If $\{V_a|A\}$ is a star-finite open covering, and if each $\mathfrak{U}_a$ is normal, then $\wedge\{V_a \circ \mathfrak{U}_a|A\}$ is normal.

Proof. Since $\mathfrak{U}_a$ is normal, there exists a normal sequence $\{\mathfrak{U}_{a,n}\}$ with $\mathfrak{U}_{a,1} < \mathfrak{U}_a$. Now (2.19) $V_a \circ \mathfrak{U}_{a,n+1} \overset{*}{<} V_a \circ \mathfrak{U}_n$; hence (2.12) $\{V_a \circ \mathfrak{U}_{a,n+1} | A\} \overset{*}{<} \wedge\{V_a \circ \mathfrak{U}_{a,n} | A\}$. Since all these coverings are open (5.1), we see that $\wedge \{ V_a \circ \mathfrak{U}_{a,n} | A$ is a normal sequence. Since $\wedge \{ V_a \circ \mathfrak{U}_{a,1} | A\} < \{V_a \circ \mathfrak{U}_a | A$, the lemma is proved.

5.3 Theorem. A star-finite covering $\mathfrak{U}$ is normal if and only if each of its associated binary coverings is normal.

Proof. Let $\mathfrak{U} = \{U_a | A\}$; then $\mathfrak{U} < \mathfrak{U}_{(a)}$ for each a. If $\mathfrak{U}$ is normal, then each $\mathfrak{U}_{(a)}$ is normal.

Suppose that each $\mathfrak{U}_{(a)}$ is normal. Let $V_a = S(U_a, \mathfrak{U})$; then $U_a \cap S(X-V_a, \mathfrak{U}) = \emptyset$, whence $V_a \circ \mathfrak{U}_{(a)} \sim \mathfrak{U}_{(a)}$ Hence (2.5, 2.3, 2.4) $\mathfrak{U} \sim \wedge\{\mathfrak{U}_{(a)}|A\} \sim \wedge\{V_a \circ \mathfrak{U}_{(a)} | A\}$. Now (2.22) $\mathfrak{U}^* = \{V_a | A\}$ is star-finite, so that (5.2) $\wedge\{V_a \circ \mathfrak{U}_{(a)} | A\}$ is normal. Hence $\mathfrak{U}$ is normal.

5.4 Theorem. A space is normal if and only if every star-finite (open) covering is normal.

Proof. Normality of the space implies (4.3, 5.3) that every star-finite (open) covering is normal. Since every binary covering is star-finite, the remainder of the theorem follows from 4.3.

6. Typical Refinement.

6.1 Theorem. If $\mathfrak{U}$ is a star-finite normal covering, then it has a star-finite and normal star refinement $\mathfrak{U}'$, such that $\mathfrak{U} \vee \mathfrak{U}'$ is star-finite.

Proof. Let $\mathfrak{U}_{(a)}$ and V_a be the same as in the proof of 5.3. Then (5.3) each $\mathfrak{U}_{(a)}$ has a normal star refinement $\mathfrak{U}_{a,1}$, which (4.2, 2.16) may be taken to be a finite covering. As in the proof of 5.2, we see that $\mathfrak{U}' = \wedge \{ V_a \circ \mathfrak{U}_{a,1}\}$ is a star-refinement of $\mathfrak{U}$ and that $\mathfrak{U}'$ is normal. Fix a', then, since $\mathfrak{B} = \mathfrak{U}^*$ is star-finite, there is a finite set α of a such that V_a meets $V_{a'}$. Let $R = \wedge\{R_a|A\}$ be a set of $\mathfrak{U}'$ which meets $U_{a'}$, where $R_a \in V_a \circ \mathfrak{U}_{a,1}$. If $a \notin \alpha$, then $V_a \cap V_{a'} = \emptyset$, so that $R_a \in V_a \circ \mathfrak{U}_{a,1}$ must be $S(X - V_a, \mathfrak{U}_{a,1})$. Hence the different R's meeting $U_{a'}$ differ only in the R_a's for $a \notin \alpha$. Since $V_a \circ \mathfrak{U}_{a,1}$ is finite for each a, it follows that $U_{a'}$ meets only a finite number of sets of $\mathfrak{U}'$, and, since $\mathfrak{U}' < \mathfrak{U}$, the theorem follows.

6.2 **. Every star-finite normal covering is the first member of a normal sequence which is a star-finite collection of coverings.**

<u>7. Special pseudo-écarts and pseudo-metrics</u>. A real valued function of two points in X is a <u>special pseudo-écart</u> or <u>spé</u> if

1° $f(x,x) = 0$,

2° $f(x,y) = f(y,x) \geq 0$.

3° for every $e > 0$, $f(x,y) < e$ and $f(y,z) < e$ imply $f(x,z) < 2e$,

4° for every $x \in X$ and $e > 0$, $\{y|f(x,y) < e\}$ is an open set.

The next lemma is a trivial generalization of a simple and important lemma due to Frink; the reader is referred to Frink 1937 (p. 134) for the simple proof.

7.1 Lemma. If f is a spé, and if $x_0, x_1, \cdots, x_{n+1}$ are any n+2 points of X, then

$$f(x_0, x_{n+1}) \leq 2f(x_0, x_1) + 4f(x_1, x_2) + \cdots + 4f(x_{n-1}, x_n) + 2f(x_n, x_{n+1}).$$

In particular,

$$f(x_0, x_{n+1}) \leq 4 \sum_0^n f(x_r, x_{r+1}).$$

If we replace 3° by the stronger condition

3^* $f(x,z) \leq f(x,y) + f(y,z)$

then a function satisfying 1°, 2°, 3^*, and 4° is a <u>pseudo-metric</u>.

7.2 **. A pseudo-metric is continuous in both variables together. (That is, given x', x" and $e > 0$, there are open sets $U' \ni x'$ and $U'' \ni x''$ such that, if $x^* \in U'$ and $x^{**} \in U''$, then $|f(x', x'') - f(x^*, x^{**})| < e$).**

7.3 Lemma. (Frink 1937). If f is a spé, then there is a pseudometric r such that

$$\tfrac{1}{4} f(x,y) \leq r(x,y) \leq f(x,y).$$

Proof. Let $r(x,y) = \inf\{ \sum_{0}^{n} f(x_r, x_{r+1})\}$, where $x_0 = x$, $x_{n+1} = y$ and the infimum is taken over all $n \geq 0$ and sets $x_1, \cdots, x_n$ of points of X . The reader may easily verify that r is a pseudo-metric, and that (7.1) it has the desired relation to f .

If f is a spé, then we may use it to define a sequence of (open) coverings. The covering $\mathfrak{U}_n$ consists of the sets $\{y | f(x,y) < 4^{-n}\}$, where $x \in X$. This sequence of coverings and any equivalent sequence is said to be associated with f .

7.4 Theorem. Every normal sequence of coverings is associated with a pseudo-metric, and any sequence of open coverings associated associated with a pseudo-metric consists of normal coverings and contains a normal subsequence.

Proof. Let $\mathfrak{U}_n$ be a normal sequence of coverings. Set $f(x,y) = 0$ if $x \in S(y, \mathfrak{U}_n)$ for all n ; set $f(x,y) = 1$ if $x \in S(y, \mathfrak{U}_1)$; and set $f(x,y) = 2^{-n}$ if $x \in S(y, \mathfrak{U}_n)$ and $x \in S(y, \mathfrak{U}_{n+1})$. Then f is clearly a spé, and $\{\mathfrak{U}_n\}$ is associated both with f and with the pseudo-metric connected with f by 7.3. We observe that

7.5 $x \in S(y, \mathfrak{U}_n)$ **implies** $r(x,y) > 2^{-(n+2)}$.

The proof of the remainder of the theorem is left to the reader.

8. Metrization. A metric is a pseudo-metric satisfying

1^* $f(x,y) = 0$ is equivalent to $x = y$,

and

5° $f(x_n, x) \longrightarrow 0$ implies $x_n \longrightarrow x$. We remark that the continuity of f (7.2) implies the converse of 5° .

The sphere of radius e (in the particular metric f) about x is defined by

8.1 $S(x,e) = \{x' | f(x,x') < e\}$.

The spherical covering of radius e is defined by

8.2 $\mathfrak{U}(e) = \{S(x,e) | x \in X\}$.

We clearly have (3*)

8.3 **.** $S(x,e) \subset S(x, \mathfrak{U}(e)) \subset S(x, 2e)$.

.We easily see that

8.4 **. A spé satisfies 5° if and only if one of its associated sequences,** $\mathfrak{U}_n$, **satisfies**

8.5 For all x, $\{S(x, \mathfrak{U}_n) | n \in N\}$ **is a nbd basis at** x.

Now we have,

8.6 Theorem. In a T_1-space the following conditions are equivalent,

8.7 X is metrizable.

8.8 There exists a countable collection $\{\mathfrak{U}_n\}$ of normal coverings satisfying 8.5.

8.9 There exists a normal sequence $\{\mathfrak{U}_n\}$ of coverings satisfying 8.5.

8.10 There exists a sequence $\{\mathfrak{U}_n\}$ of coverings satisfying 3.4 and 8.5

All these conditions are implied by

8.11 X is a normal space with a countable basis.

Proof. If 8.7 holds, then (7.4) the metric is associated with a normal sequence, which (8.4) satisfies 8.5; hence 8.9 holds. Conversely, if 8.9 holds, then (7.4) there is an associated pseudo-metric, which (8.4) satisfies 5°; hence 8.7 holds.

If 8.8 holds, then there exist normal coverings $\mathfrak{V}_n$ so that $\mathfrak{V}_{n+1} {}^* < \mathfrak{V}_n \wedge \mathfrak{U}_n$, for $\mathfrak{V}_n$, $\mathfrak{U}_n$ and therefore $\mathfrak{V}_n \wedge \mathfrak{U}_n$ are normal. It is clear that $\{\mathfrak{V}_n\}$ is a normal sequence, and (III-6.17) satisfies 8.5; hence 8.9 holds. Clearly 8.9 implies 8.8.

Now 8.9 clearly implies 8.10. If 8.10 holds, then (3.3) $\{\mathfrak{U}_n^\Delta\}$ are normal and (III-6.17, 3.5) satisfy 8.5; hence 8.9 holds.

If 8.11 holds, then (4.7) the countable collection of basic binary coverings satisfies (8.5), so that 8.8 holds.

8.12 Theorem. Every (open) covering of a metrizable space is normal.

Proof. Let X be the metrizable space, and let f be some metric for X . Let $\mathfrak{U} = \{U_a | A\}$ be an open covering of X . Now (8.2) for each x , some e > 0 , and some a , $x \in S(x,e) \subset U_a$. We choose $e(x|X)$ and $a(x|X)$ so that

$$0 < e(x) < 1 \ ; \quad S(x, 4e(x)) \subset U_{a(x)} \ .$$

Then $\mathfrak{V} = \{V_x | X\}$, where $V_x = S(x, e(x))$, is an open covering of X . We shall prove that $\mathfrak{V} \lhd \mathfrak{U}$

Fix x' , and consider $H = \{x | x' \in V_x\}$. ($x' \in H$, so that $H \neq \emptyset$) . We may choose $x^* \in H$, so that

$$e(x^*) > \frac{2}{3}\sup\{e(x) \mid x \in H\}.$$

Then, if $x \in H$,

$$V_x = S(x,e(x)) \subset S(x',2e(x)) \subset S(x',3e(x^*)) \subset S(x^*,4e(x^*)).$$

This holds for any x such that $x' \in V_x$; hence

$$S(x',\mathfrak{V}) \subset S(x^*,4e(x^*)) \subset U_{a(x^*)}$$

Since x' was arbitrary, we have shown $\mathfrak{V} \lhd \mathfrak{U}$

If $\mathfrak{U}_1$ is any open covering of X we have just proved that we can find successively coverings $\mathfrak{U}_n$ so that $\mathfrak{U}_{n+1} \lhd \mathfrak{U}_n$. Hence (3.3) $\{\mathfrak{U}_{2n}\}$ is a normal sequence; and, since $\mathfrak{U}_2 < \mathfrak{U}_1$, $\mathfrak{U}_1$ is normal.

An immediate consequence of this theorem is

8.13 **. Every metrizable space is normal.**

We shall say that a space is <u>fully normal</u> if every open covering is normal. We may rephrase 8.12 as

8.14 **. Every metrizable space is fully normal.**

<u>9. Continuous Real Values Functions</u>. If we have a pseudo-metric r, we define a corresponding function of subsets by

$$r(x,K) = \inf\{r(x,k) \mid k \in K\}$$

and

$$r(H,K) = \inf\{r(h,K) \mid h \in H\}.$$

9.1 **. $r(x,H)$ is continucus in x.**

9.2 Theorem. If $\mathfrak{U}$ is a normal covering and H is any set, then there exists a real-valued continuous function h so that

$$h(x) = \begin{cases} 0, & x \in H, \\ 1, & x \in S(H,\mathfrak{U}). \end{cases}$$

<u>Proof</u>. Let $\{\mathfrak{U}_n\}$ be a normal sequence, where $\mathfrak{U}_1 < \mathfrak{U}$; let r be the associated pseudo-metric determined in 7.5. If $x \notin S(H,\mathfrak{U})$, then (7.6) $r(x,H) \geq 1/4$. So we may take

$$h(x) = \min\{1, 4r(x,H)\},$$

and thus prove the theorem.

We now generalize Urysohn's Lemma.

9.3 Theorem. The binary covering $\{U',U''\}$ is normal if and only if there exists a continuous real valued function h such that

$$h(x) = \begin{cases} 0, & x \notin U' \\ 1, & x \notin U'' \end{cases}$$

Proof. If such an h exists, $|h(x)-h(y)|$ is a pseudo-metric associated (7.4) with a normal sequence $\mathfrak{U}_n$. Since $\mathfrak{U}_k < \{U', U''\}$ for some k, it follows that $\{U', U''\}$ is normal.

If $\mathfrak{U} = \{U', U''\}$ is normal, then, since $S(X-U', \mathfrak{U}) = U''$, we may apply 9.2 to $X-U'$ and $\mathfrak{U}$.

10. Historical remarks. The notions of $*<$, $\triangleleft$, and the related ideas appear (in an incompletely formalized state) in topology since Urysohn's earliest work on the metrization problem. How much further back they may be traced I do not know. However, I believe that this is the first attempt at their systematic exploitation.

Chapter VI

STRUCTS

1. Introduction.
2. Uniformities.
3. Agreement.
4. Structs.
5. Uniform Continuity.
6. The Metric Case.
7. Analogy.
8. Completeness.
9. Completion.
10. Historical Remarks.

1. Introduction. In this chapter we are concerned with structs. A struct is a space in which there is a suitable notion of uniformity. (The connection between the notion of a struct and some of the similar notions recently considered is discussed in §10). We develop only the fundamentals of the theory.

In §2 we define uniformity, and in §3 we discuss the agreement of uniformity and topology. In §4 we discuss structs, and in §5 uniform continuity. In §6 we discuss the special case of metric spaces, and in §7 we note some interesting analogies between the transitions from set to space and from space to struct. In §8 we define and discuss completeness, and in §9 we carry thru the completion of a struct.

2. Uniformities. A uniformity in a space X is a collection $\{\mathfrak{U}\}$ of (open) coverings of X satisfying

2.1 If $\mathfrak{U} < \mathfrak{U}'$ and $\mathfrak{U} \in \{\mathfrak{U}\}$, then $\mathfrak{U}' \in \{\mathfrak{U}\}$.

and

2.2 If $\mathfrak{U}', \mathfrak{U}'' \in \{\mathfrak{U}\}$, then, for some $\mathfrak{U}''' \in \{\mathfrak{U}\}$, $\mathfrak{U}''' \overset{*}{<} \mathfrak{U}'$, $\mathfrak{U}''' \overset{*}{<} \mathfrak{U}''$.

A subcollection $\{\mathfrak{V}\} \subset \{\mathfrak{U}\}$ is a basis for the uniformity $\{\mathfrak{U}\}$ if

2.3 $\{\mathfrak{U}\} = \{\mathfrak{U} \mid \mathfrak{V} < \mathfrak{U}, \mathfrak{V} \in \{\mathfrak{V}\}\}$

We say that $\{\mathfrak{V}\}$ induces the uniformity $\{\mathfrak{U}\}$. A collection is a basis for some uniformity if and only if it satisfies 2.2.

A subcollection $\{\mathfrak{W}_a | A\} \subset \{\mathfrak{U}\}$ is a sub-basis for the uniformity $\{\mathfrak{U}\}$ if

2.5 $\{\mathfrak{U}\} = \{\mathfrak{U} \mid \bigwedge_{\alpha} \mathfrak{W}_a < \mathfrak{U}, \ \alpha \subset A\}$.

We easily see that

2.6 **. a) A collection $\{\mathfrak{W}_a | A\}$ is a sub-basis for some uniformity if and only if it satisfies**

2.7 If $\mathfrak{W}' < \{\mathfrak{W}\}$, then, for some $\mathfrak{W}'' \in \{\mathfrak{W}\}$, $\mathfrak{W}'' \ {}^{*}{<} \ \mathfrak{W}'$.

b) A collection $\{\mathfrak{V}\}$ is a basis for some uniformity if and only if it satisfies

2.8 If $\mathfrak{V}', \mathfrak{V}'' \in \{\mathfrak{V}\}$, then, for some $\mathfrak{V}''' \in \{\mathfrak{V}\}$, $\mathfrak{V}''' < \mathfrak{V}' \wedge \mathfrak{V}''$.

and

2.9 If $\mathfrak{V}' \in \{\mathfrak{V}\}$, then, for some $\mathfrak{V}'' \in \{\mathfrak{V}\}$, $\mathfrak{V}'' \lhd \mathfrak{V}'$.

We also see that

2.10 **. Two collections of coverings, at least one of which is a basis of some uniformity, are bases of the same uniformity if and only if they are equivalent (V-2).**

We shall often have occasion to use

2.11 **. If $\{\mathfrak{V}_a | A\}$ is a basis for a uniformity to which $\mathfrak{U}$ belongs, and if n is any integer, then, for some a A, we have**

$$\mathfrak{V}_a^{*\cdots*(n\ \text{times})} \ {}^{*}{<} \ \mathfrak{U},$$

$$\{S^n(x, \mathfrak{V}_a) | X\} \ {}^{*}{<} \ \mathfrak{U},$$

and, for all $H \subset X$,

$$S^n(H, \mathfrak{U}_a) \subset S(H, \mathfrak{U}).$$

There is a notable connection between normality and uniformity, for we have,

2.12 Theorem. Every covering appearing in a uniformity is normal.

The collection of all the normal coverings (of any T-space) is a uniformity.

Proof. Let $\{\mathfrak{U}\}$ be a uniformity, and let $\mathfrak{U}_1 \in \{\mathfrak{U}\}$. Then (2.2) we may choose $\mathfrak{U}_n \in \{\mathfrak{U}\}$ so that $\mathfrak{U}_{n+1}{}^* < \mathfrak{U}_n$. is a normal sequence, and $\mathfrak{U}_1 < \mathfrak{U}_1$, hence $\mathfrak{U}_1$ is normal.

The second statement follows from the definition of normality and V-3.1.

If $\{\mathfrak{U}\}$ and $\{\mathfrak{V}\}$ are uniformities of X, then we say that $\{\mathfrak{U}\}$ is <u>finer than</u> $\{\mathfrak{V}\}$, if $\{\mathfrak{U}\} \supset \{\mathfrak{V}\}$.

We easily see (2.6 b), V-4) that the finite normal coverings of X form a basis for a uniformity of X. We denote this uniformity by fX. We denote the uniformity made up of all normal coverings by aX. If the enumerable normal coverings are a basis for a uniformity, then we denote the uniformity by eX. It is important to observe that these uniformities are topological invariants of X.

We may restate the last part of 2.12 as

2.13 **. aX is the finest uniformity of X.**

3. Agreement. We begin with

3.1 Lemma. If $\{\mathfrak{U}\}$ is a uniformity of which $\{\mathfrak{U}_a | A\}$ is a basis and $\{\mathfrak{U}_b | B\}$ is a sub-basis; then the following conditions are equivalent.

3.2 For all x, $\{S(x,\mathfrak{U}) | \mathfrak{U} \in \{\mathfrak{U}\}\} = \{U | x \in U, U \text{ is open}\}$.

3.3 For all x, $\{S(x,\mathfrak{U}_a) | A\}$ is a nbd basis at x.

3.4 For all x, $\{S(x,\mathfrak{U}_b) | B\}$ is a nbd sub-basis at x.

Proof. Assume 3.2, and let N be a nbd of x. Then $x \in U \subset N$, and, for some $\mathfrak{U} \in \{\mathfrak{U}\}$, $U = S(x,\mathfrak{U})$. Let $\mathfrak{U}_a < \mathfrak{U}$: then $S(x,\mathfrak{U}_a) \subset S(x,\mathfrak{U}) \subset N$; hence 3.3 holds. Let β be such that $\bigwedge_\beta \mathfrak{U}_b < \mathfrak{U}$. Then (V-2.25) $\bigcap_\beta S(x,\mathfrak{U}_b) \subset S(x,\mathfrak{U}) \subset N$, so that 3.4 holds. The converse arguments are the same, with the addition of the fact that, if $S(x,\mathfrak{U}) \subset U$ and $\mathfrak{U}' = \mathfrak{U} \cup \{U\}$, then $S(x,\mathfrak{U}') = U$.

If these equivalent conditions are satisfied, then we say that the uniformity <u>agrees</u> with the topology. It is clear that if $\{\mathfrak{U}\}$ agrees with the topology of X, then so does every finer uniformity. Not every T-space possess a uniformity agreeing with the topology. In fact we have,

3.5 Theorem. The following conditions on a T_1-space are equivalent:

3.6 If $x' \in U'$ and U' is open, then there is a continuous real-valued function h with

$$h(x) = \begin{cases} 0, & x = x', \\ 1, & x \notin U'. \end{cases}$$

3.7 If $x' \in U'$ and U' is open, then $\{X - \{x'\}, U'\}$ is a normal covering.

3.8 Some uniformity gX of X agrees with the topology of X.

3.9 The uniformity aX agrees with the topology of X.

3.10 The uniformity fX agrees with the topology of X.

Proof: We know (V-9.3) that 3.6 and 3.7 are equivalent, since x' is closed, as a point of a T_1-space. Assume 3.7, and let U be a nbd of x. Then (3.7) $\{X - \{x\}, U\}$ is normal and belongs to fX. Since the star of x in this covering is U, 3.10 holds. 3.10 clearly implies 3.8. 3.8 implies 3.9, since (2.13) aX is finer than gX. We need now only prove that 3.9 implies 3.7. Assume 3.9, and let $x' \in U'$. Then, for some $\mathfrak{U} \in aX$, $S(x', \mathfrak{U}) \subset U'$. This means that $\mathfrak{U} < \{X - \{x'\}, U'\}$, and so $\{X - \{x'\}, U'\}$ is normal and 3.7 holds.

A T_1-space satisfying these equivalent conditions is a completely regular space (Tychonoff). We may derive several consequences from 3.5. We begin with

3.11 Lemma. Every normal T_1-space is completely regular.

Proof. $\{X - \{x\}, U\}$ is a binary open covering and hence normal.

3.12 Remark. For T_1-spaces we may regard 3.7 as a natural dilution of the requirement of normality (when this requirement is expressed as "every binary open covering is normal.")

3.13 Lemma. If a T_1-space X satisfies any of 3.6 through 3.10 then so do its subspaces.

Proof. This is clear (III-5.2, III-7.3) for 3.6, hence (3.5) it is true for all.

For the remainder of this chapter we will consider only completely regular spaces. Hence we will use "space" to mean "completely regular space."

3.14 Definition. A *struct* is a (completely regular) space and a uniformity of that space which agrees with the topology of the space.

We refer to the struct, loosely, in the same way that we refer to the uniformity. Thus aX is a struct as well as a uniformity. We use gX , hX , ... for general structs over the space X .

4. Structs. A basis [sub-basis] for the uniformity of a struct is a basis [sub-basis] for the struct. A basis consisting of finite coverings is an f-basis.

The coverings belonging to the uniformity of a struct gX are said to be large in gX .

If Y is a subspace of X , and gX is a struct over X , then we may define a struct gY by

4.1 $\mathfrak{V}$ is large in gY if $\mathfrak{V}=\mathfrak{U}\cap Y$, where $\mathfrak{U}$ is large in gX.

We call gY a substruct of gX if 4.1 holds. It is important to notice that if Y is a subspace of X it need not be true that aY is a substruct of aX , or that fY is a substruct of fX . And it may happen that Y is a subspace of X and gY is a struct, but there is no struct hX of which gY is a substruct.

We often deal with a struct gX and a fixed basis $\{\mathfrak{U}(a)|A\}$ of gX . Under these conditions, we often abbreviate $S(H,\mathfrak{U}(a))$ to $S(H,a)$. If gY is a substruct of gX , and if $\{\mathfrak{U}(a)|A\}$ is a basis of gX , then $\{Y\cap\mathfrak{U}(a)|A\}$ is a basis of gY .

We see that we have,

4.2 **. A necessary and sufficient condition for $x(\beta|B)$ to converge to x, is the existence of an integer n and sets $\{\beta_a|A\}$ such that $\beta\supset\beta_a$ implies $x(\beta)\in S^n(x,a)$. (Here $\{\mathfrak{U}(a)|A\}$ is a basis for gX.)**

A result of considerable importance is

4.3 Lemma. If X is compact, if gX is a struct over X with basis $\{\mathfrak{U}(a)|A\}$, and if $\mathfrak{V}$ is any (open) covering of X; then, for some a, $\mathfrak{U}(a)\lhd\mathfrak{V}$.

Proof. We choose $a(\alpha|A)$ so that $\mathfrak{U}(a(\alpha)) < \mathfrak{U}(a)$ for all a in α . If the lemma were false, then, for each α and some $x(\alpha)$, $S(x(\alpha),a(\alpha))$ would fail to lie on a single set of $\mathfrak{V}$ X is compact, so (IV-4.3) $x(\alpha|A)$ has a cluster point x' , which must be in an open set $V'\in\mathfrak{V}$. Now (3.3), for some a' , $S(x',a')\subset V'$. And (2.11) we may choose a'' so that $S^2(x',a''\subset S(x',a')$. Take α' so that $\alpha'\ni a''$ and $x(\alpha')\in S(x',a')$. Then $S(x(\alpha'),a(\alpha'))\subset S(x(\alpha'),a'')$

$\subset S^2(x',a'') \subset S(x',a') \subset V'$, which is a contradiction.

From this lemma we derive

4.4 Theorem. If X is compact (and completely regular), then it has one and only one uniformity agreeing with its topology, and this uniformity is made up of all (open) coverings.

On a compact (and completely regular space) there is one and only one struct.

Every (open) covering of a compact completely regular space is normal.

Proof. Some gX is (3.8) a struct over X and (4.3) every (open) covering of X is large in gX . Hence gX = aX for all gX . Furthermore, all (open) coverings belong to aX .

An easy consequence of this result is

4.5 **. Every compact (completely regular) space is fully normal.**

We call the unique struct over a compact space a <u>compact struct</u>, and often refer to it by X alone.

From V-4.6 we easily see that

4.6 **. If $\{\mathfrak{U}(a)|A\}$ is a basis for the compact struct X, and if F' and F" are closed disjoint sets in X, then, for some a, $F' \cap S(F'',a) = \emptyset$**

We say that gX is largely compact if it has an f-basis, Clearly every compact struct is <u>largely compact</u>. The name is justified by

4.7 Lemma. A struct gX is largely compact if and only if every large covering has a finite subcovering.

Proof. Let gX be largely compact, let $\mathfrak{V}$ be large in gX , and let $\{\mathfrak{U}(a)|A\}$ be an f-basis for gX . For some a , $\mathfrak{U}(a) < \mathfrak{V}$. Now $\mathfrak{U}(a) = U_1,\cdots,U_n$, and there are $V_i \in \mathfrak{V}$ with $U_i \subset V_i$ for each i . Let $\mathfrak{V}' = \{V_1,\cdots,V_n\}$; since $\bigcup\{V_i\} \supset \bigcup\{U_i\} = X$, $\mathfrak{V}'$ is the desired finite covering.

Conversely, if every large covering has a finite subcovering, these finite subcoverings form an f-basis for gX .

5. <u>Uniform Continuity</u>. We have

5.1 Lemma. If φ is a continuous function on gX to hY, whose bases are $\{\mathfrak{U}(a)|A\}$ and $\{\mathfrak{V}(b)|B\}$; then the following conditions are equivalent.

5.2 If $\mathfrak{B}$ is large, then $\varphi^{-1}(\mathfrak{B})$ is large.

5.3 There is a function $a(b|B)$, such that $x' \in S(x'', a(b))$ implies $\varphi(x') \in S(\varphi(x''), b)$.

5.4 $\varphi^{-1}(\mathfrak{B}(b))$ is large for all b.

Proof. Assume 5.4, then $\varphi^{-1}(\mathfrak{B}(b))$ is large, so that, for some $a = a(b)$, $\mathfrak{B}(a) < \varphi^{-1}(\mathfrak{B}(b))$. This clearly implies 5.3.

Assume 5.3, and let $\mathfrak{B}$ be large. For some b, $\mathfrak{B}(b) < \mathfrak{B}$; hence $\varphi^{-1}(\mathfrak{B}(b)) < \varphi^{-1}(\mathfrak{B})$. Choose b' so that $\mathfrak{B}(b')^* < \mathfrak{B}(b)$, and choose a' so that $\mathfrak{U}(a')^* < \mathfrak{U}(a(b'))$; if we prove $\mathfrak{U}(a') < \varphi^{-1}(\mathfrak{B}(b))$, we will have proved 5.2. Let $U \in \mathfrak{U}(a')$ and $x' \in U$. Then $U \subset S(x', a')$, and $\varphi(U) \subset \varphi(S(x', a')) \subset S(x'), b') \subset V$, where $V \in \mathfrak{B}(b)$. Therefore $U \in \varphi^{-1}(V)$ and 5.2 holds.

Clearly 5.2 implies 5.4.

If φ satisfies these conditions, we say that it is uniformly continuous.

The following result is interesting and useful.

5.5 Theorem. A continuous function on an f-struct fX to a largely compact struct gY is uniformly continuous.

A continuous function on an a-struct aX to a struct gY is uniformly continuous.

Proof. These results follow immediately from V-3.7 and 5.4.

Combining this result with 4.6 we have

5.6 **. Every continuous function on a compact struct is uniformly continuous.**

A one-to-one correspondence between two structs, which is uniformly continuous in both senses, is a unimorphism. From the point of view of the theory of structs, two unimorphic structs are abstractly the same.

6. The Metric Case. The intuitive nature of the machinery which we have set up may perhaps be clarified by considering the special case of metric spaces. We begin with

6.1 Theorem. A space is metrizable if and only if it is the space of a struct with a countable basis.

Proof. This is essentially a rewording of part of V-8.6, for V-8.8 holds in a struct with a countable basis, and V-8.9 implies the existence of a struct with a countable basis (the normal sequence).

If we have a metric space, that is, a metrizable space with a particular metric; then on that space we can erect a struct in a natural way. We shall use the notation of V-8. The coverings, $\mathfrak{U}_e$, by e-spheres are a basis for the struct. That is, a covering is large if, for some $e > 0$ and all x , the set $S(x,e)$ is in a single set of the covering. In the case of a compact metric space, the greatest such e is called the Lebesgue constant of the open covering. 4.3 is a generalization of the classical result that any open covering of a compact metric space has a positive Lebesgue constant. From 4.6 we can derive the corresponding result for closed coverings.

We see that our definition of uniform continuity reduces in the case of a metric space to the classical one of analysis

The notion of "largely compact" corresponds to that of "total-boundedness," and 3.10 is related to the classical result that every separable metrizable space can be meterized so that it is totally bounded.

How much of the individuality of the metric are we using? It is easy to see that two metrics yield the same struct if and only if they uniformly continuous functions of each other.

7. Analogy. There is an interesting and unexplained analogy between structs and T_1-spaces. We give parallel definitions and theorems in parallel columns.

A set.	A completely regular space.
A T -space.	A struct.
A subset.	A normal covering.
An open subset.	A large covering.
A discrete space (every subset is open).	An a-struct (every normal covering is large).
Every function on a discrete space is continuous.	Every continuous function on an a-struct is uniformly continuous.
A finite set.	A compact space.
A finite set has only one topology (as a T_1-space).	A compact space has only one uniformity (as a struct).

8. Completeness. One important property of metric spaces which we have not considered for structs is completeness.

We begin with the notion of Cauchy mapping. A mapping

$x(b|\mathfrak{B})$ of a directed system $\mathfrak{B}$ into a struct gX is a Cauchy mapping if it satisfies

8.1 If $\mathfrak{U}$ is large, then, for some x', $x(b|\mathfrak{B})$ is ultimately in $S(x',\mathfrak{U})$.

It is clearly sufficient that 8.1 hold when $\mathfrak{U}$ runs thru a basis of gX. This is easily seen to generalize the usual notion of a Cauchy sequence in a metric space. We shall most frequently be concerned with Cauchy phalanxes.

We see immediately (Most of the proofs left to the reader in this § require frequent use of 2.11) that

8.2 **. Every convergent phalanx is a Cauchy phalanx.**

8.3 **. A cluster point of a Cauchy phalanx is a limit of the phalanx.**

8.4 **. The uniformly continuous image of a Cauchy phalanx is a Cauchy phalanx.**

8.5 **. If gX is a substruct of gY, and if $x(\Gamma)\subset X$, then $x(\gamma|C)$ is Cauchy in gX if and only if it is Cauchy in gY.**

Two Cauchy phalanxes are a Cauchy pair if the result of meshing them (III-2) is a Cauchy phalanx.

8.6 **. If $\{\mathfrak{U}_a|A\}$ is a basis for gX, then**

a) $x'(\beta|B)$ and $x''(\gamma|C)$ are a Cauchy pair if there exists an integer n and points $\{x_a|A\}$, so that both $x'(\beta|B)$ and $x''(\gamma|C)$ are ultimately in $S^n(x_a,a)$, for each a.

b) If $x'(\beta|B)$ and $x''(\gamma|C)$ are a Cauchy pair, there exist points $\{x_a|A\}$ so that $x'(\beta|B)$ and $x''(\gamma|C)$ are both ultimately in $S(x_a,a)$ for each a.

8.7 **. The uniformly continuous image of a Cauchy pair is a Cauchy pair.**

8.8 **. If gX is a substruct of gY, and $x'(\beta|B)$ and $x''(\gamma|C)$ are in X, then they are a Cauchy pair in gX if and only if they are a Cauchy pair in gY.**

8.9 **. The relation of being a Cauchy pair is an equivalence relation.**

8.10 **. If $x'(\beta|B)$ and $x''(\gamma|C)$ are a Cauchy pair, and if $x'(\beta|B)$ converges to x', then so does $x''(\gamma|C)$. Conversely, any pair of phalanxes converging to x is a Cauchy pair.**

A struct gX is complete if every Cauchy phalanx is convergent. From 8.4 we easily see that

8.11 **. A unimorphic image of a complete struct is a complete struct.**

From 8.2 we see that

8.12 **. If gX is complete and gY is a substruct of gX; then gY is complete if and only if Y is closed in X.**

8.13 Lemma. In a largely compact struct each ultraphalanx is a Cauchy phalanx.

Proof. An ultraphalanx is decided about any finite covering. Since X has an f-basis, we see that every ultraphalanx in gX is a Cauchy phalanx.

8.14 Theorem. A largely compact struct is complete if and only if it is compact.

A compact struct is complete

Proof. If gX is compact, then every Cauchy phalanx has a cluster point, which is (8.3) a limit. Hence gX is complete.

If gX is complete and largely compact, then every ultraphalanx, being (8.13) a Cauchy phalanx, is convergent. Hence gX is compact.

We say that hY is a completion of gX, if there is a unimorphism φ, such that $\varphi(gX)$ is a substruct of hY, and $\overline{\varphi(gX)} = hY$.

8.15 Lemma. If gX is a substruct of gY, and φ is a uniformly continuous function on gX with values in a complete struct hZ; then there is a unique extension ψ of φ to $\overline{gX}$ and has values in hZ.

Proof. Let $y \in \overline{X}$, then (8.2) there is a Cauchy phalanx $x(\gamma | C)$ converging to y, with $x(\Gamma) \subset X$. Now (8.4, 8.5) $\varphi(x(\gamma)) | C)$ is a Cauchy phalanx in hZ, and hence has a limit z in hZ. If ψ is a continuous extension of φ, then (III-7.3) $\varphi(y) = z$. We see that this process leads to a unique z for each y. For if two different phalanxes converge to y, they are (8.10) a Cauchy pair, and (8.7, 8.8) their images are also a Cauchy pair and (8.10) converge to the same z. Thus the only possible continuous extension ψ of φ is determined by this process. We must show that it is uniformly continuous.

Let $\{\mathfrak{U}(a)|A\}$ be a basis for gX, and let $\{\mathfrak{V}(b)|B\}$ be a basis for $h\varphi(\overline{X})$. Then $\{X\cap\mathfrak{U}(a)|A\}$ is a basis for gX. Since φ is uniformly continuous, there is (5.3) a function $a(b|B)$ such that $x'\in S(x'',a(b))$ implies $\varphi(x')\in S(\varphi(x'',b)$. Choose $b'(b|B)$ so that $\mathfrak{V}(b'(b))^*<\mathfrak{V}(b)$, and let $a'(b) = a(b'(b))$. Suppose that y' and y'' belong to X and that $y'\in S(y'',a'(b))$. This means that there is an open set $U\in\mathfrak{U}(a'(b))$, with $y', y''\in U$. Now (III-5.5) $y', y''\in\overline{X\cap U}$, so there are (8.15) phalanxes $x'(\gamma|C)$ converging to y' and $x''(\gamma|C)$ converging to y'', with $x'(\Gamma)\subset X\cap U\supset x''(\Gamma)$. Since $x'(\gamma)\in S(x'(\emptyset),a'(b))$ and $x''(\gamma)\in S(x'(\emptyset),a'(b))$ for all γ, we see that, for all γ, both $\varphi(x'(\gamma))$ and $\varphi(x''(\gamma))$ belong to $S(\varphi(x'(\emptyset),b'(b))$, which is contained in some set V of $\mathfrak{V}(b)$. Since, for all γ, $\varphi(x'(\gamma))$ and $\varphi(x''(\gamma))$ belong to V, so do the limits $\varphi(y')$ and $\varphi(y'')$ of these phalanxes. Thus $\varphi(y')\in S(\varphi(y''),b)$. We have proved (5.3) that φ is uniformly continuous.

We may prove in the same way that

8.16 **. If:- gX is a substruct of gY, hV is a substruct of hZ, $\overline{X}$=Y, $\overline{V}$=Z, gY and hZ are complete, and gX and hY are unimorphic; then gY and hZ are unimorphic**

And from this we see immediately that

8.17 ** Any two completions of a struct are unimorphic.**

<u>9. Completion</u>. We are going to construct a completion of an arbitrary struct gX. Although the result, as we know (8.17) is essentially unique, we find it convenient to use a fixed basis (of a special sort) in its construction.

9.1 Lemma. gX has a basis $\{\mathfrak{U}(\alpha)|A\}$, where $\alpha'\supset\alpha''$ implies $\mathfrak{U}(\alpha')<\mathfrak{U}(\alpha'')$.

<u>Proof</u>. Let $\{\mathfrak{U}(a)|A\}$ be any basis for gX, and let $\mathfrak{U}(\alpha) = \wedge\{\mathfrak{U}(a)|\alpha\}$; then $\{\mathfrak{U}(\alpha)|A\}$ is such a basis.

We shall constantly use such a basis for gX and abbreviate $S(H,\mathfrak{U}(\alpha))$ as $S(H,\alpha)$. An A-phalanx is of <u>rank n</u> if

9.2 $\alpha'\supset\alpha$ and $\alpha''\supset\alpha$ imply $x(\alpha')\in S^n(x(\alpha''),\alpha)$.

Obviously,

9.3 **. Every phalanx of finite rank is a Cauchy phalanx.**

The importance of phalanxes of finite rank is a consequence of

9.4 Theorem. If $x'(\beta|B)$ is a Cauchy phalanx in gX, then there exists a phalanx of rank 2, $x(\alpha|A)$, which forms a Cauchy pair with $x'(\beta|B)$.

Proof. By 8.1, there exist points $x(\alpha|A)$ so that, for each α, $x'(\beta|B)$ is ultimately on $S(x(\alpha),\alpha)$. If $\alpha' \supset \alpha$ and $\alpha'' \supset \alpha$, then (9.1) $x'(\beta|B)$ is ultimately on both $S(x(\alpha'),\alpha)$ and $S(x(\alpha''),\alpha)$; hence (IV-2.5) these sets meet, and $x(\alpha') \in S^2(x(\alpha''),\alpha)$. Therefore $x(\alpha|A)$ is of rank 2, and (8.6) $x(\alpha|A)$ and $x'(\beta|B)$ are a Cauchy pair.

From 8.10 and 9.4 we see that

9.5 **. gX is complete if and only if each phalanx of rank 2 is convergent.**

From 8.2, 8.10 and 9.4 we see that

9.6 **. The topology of X is described by the convergence of phalanxes of rank 2.**

We need,

9.7 Lemma. If $x(\alpha|A)$ is of rank r, and if, for some n and all α, $x'(\alpha|A)$ is ultimately in $S^n(x(\alpha),\alpha)$; then $x(\alpha|A)$ and $x'(\alpha|A)$ are a Cauchy pair.

Proof. $x'(\alpha|A)$ is ultimately in $S^n(x(\alpha),\alpha) \subset S^{n+r}(x(\alpha),\alpha)$. $x(\alpha|A)$ is ultimately (9.2) in $S^r(x(\alpha),\alpha) \subset S^{n+r}(x(\alpha),\alpha)$. Hence (8.6) $x'(\alpha|A)$ and $x(\alpha|A)$ are a Cauchy pair.

9.8 Lemma. If $x(\alpha|A)$ is of rank r, and if $x'(\alpha) \in S^n(x(\alpha),\alpha)$, then $x'(\alpha|A)$ is of rank 2n r, and $x(\alpha|A)$ and $x'(\alpha|A)$ are a Cauchy pair.

Proof. If $\alpha'' \supset \alpha$ and $\alpha'' \supset \alpha$, then $x'(\alpha') \in S^n(x(\alpha'),\alpha') \subset S^n(x(\alpha'),\alpha)$, $x(\alpha') \in S^r(x(\alpha''),\alpha)$, $x''(\alpha'') \in S^n(x(\alpha''),\alpha'') \subset S^n(x(\alpha''),\alpha)$; hence $x''(\alpha') \in S^{2n+r}(x(\alpha''),\alpha'')$. Thus $x'(\alpha|A)$ is of rank $2n+r$, and the remainder of the lemma follows from 9.7.

We are now prepared to proceed with the definition of the completion of gX.

We consider the equivalence classes of phalanxes of finite rank with respect to the relation (8.6) of being a Cauchy pair; they make up a set X^*. We define a mapping $H \to H^*$ of 2^X into 2^{X^*} by

9.9 $H^* = \{x^* | x(\alpha|A) \in x^*$ implies that $x(\alpha|A)$ decides for $H\}$.

It is clear that

9.10 $H \subset K$ implies $H^* \subset K^*$.

and

9.11 $(H \cap K)^*$ $H^* \cap K^*$.

We define $\{U^*|U \text{ open in } X\}$ as a basis for the open sets of X^*. It is clear (9.11) that X^* is a T-space.

We define $\mathfrak{U}^*(\alpha)$ by $\mathfrak{U}^*(\alpha) = \{U^*|U \in \mathfrak{U}(\alpha)$; and we have

9.12 Lemma. $\mathfrak{U}^*(\alpha)$ **is a covering of** X^*.

Proof. Let $x(\alpha|A) \in x^*$. Choose (2.11) α' so that $\{S^3(x,\alpha')\} < \mathfrak{U}(\alpha)$. Since $x(\alpha|A)$ is Cauchy, it is ultimately on $S(x',\alpha')$ for some x'. If $x'(\alpha|A) \in x^*$, then $x'(\alpha|A)$ and $x(\alpha|A)$ are a Cauchy pair, and (8.6) there is an x'' so that they are both ultimately on $S(x'',\alpha')$. Clearly $S(x'',\alpha')$ meets $S(x',\alpha')$, so that $x'(\alpha|A)$ is ultimately on some $U \supset S^3(x',\alpha') \supset S(x'',\alpha')$, where $U \in \mathfrak{U}(\alpha)$. Hence $x^* \in U^*$ and the lemma is proved.

Now from 9.11 it is clear that

9.13 $\alpha' \supset \alpha''$ **implies** $\mathfrak{U}^*(\alpha') < \mathfrak{U}^*(\alpha'')$.

We now have

9.14 Lemma. If $x(\alpha|A) \in x^*$ **and has rank n, then**

$$x^* \in S(x^*,\alpha) \in (S^{n+1}(x(\alpha),\alpha))^*.$$

Proof. Let $x'(\alpha|A) \in x'^* \in S(x^*,\alpha)$; then there is a set $U \in \mathfrak{U}(\alpha)$ such that x'^* and x^* both belong to U^* $x(\alpha|A)$ is ultimately in both U and $S^n(x(\alpha),\alpha)$, so that $U \subset S^{n+1}(x(\alpha),\alpha)$. Since $x'(\alpha|A)$ is ultimately in $U \subset S^{n+1}(x(\alpha),\alpha)$, we have $x'^* \in (S^{n+1}(x(\alpha),\alpha))^*$.

9.15 Lemma. If $x(\alpha|A) \in x^* \in U^*$, **and n is a given integer; then, for some** α, $S^n(x(\alpha),\alpha) \subset U$.

Proof. If the lemma were false, then, for all α, $S^n(x(\alpha),\alpha) \not\subset U$. So we might choose $x'(\alpha)$ so that $x'(\alpha) \in S^n(x(\alpha),\alpha)$ and $x'(\alpha) \notin U$. Let $x(\alpha|A)$ have rank r ; then (9.8) $x'(\alpha)$ has rank $2n+r$, while $x(\alpha|A)$ and $x'(\alpha|A)$ are a Cauchy pair; hence $x'(\alpha|A) \in x^*$. Since each $x'(\alpha) \notin U$, $x^* \notin U^*$ This contradiction proves the lemma.

9.16 Lemma. $\{\mathfrak{U}^*(\alpha)|A\}$ **satisfies 2.8, 2.9 and 3.3.**

Proof. Fix α, and take (2.11) α' so that $\{S^3(x,\alpha')|X\} < \mathfrak{U}(\alpha)$. Choose $x(\alpha|A) \in x^*$ of rank 2; then (9.14, 9.10) $S(x^*,\alpha') \subset (S^3(x(\alpha'),\alpha'))^* \subset U^*$, where $U \in \mathfrak{U}(\alpha)$. Hence $\mathfrak{U}^*(\alpha') \triangleleft \mathfrak{U}^*(\alpha)$, and 2.9 holds. 2.8 is an easy consequence of 9.13.

Let $x^* \in U^*$, and let $x(\alpha|A) \in x^*$ be of rank n ; then (9.15) there is an α such that $S^{n+1}(x(\alpha),\alpha) \subset U$. Now (9.14) $S(x^*,\alpha) \subset (S^{n+1}(x(\alpha),\alpha))^* \subset U^*$; hence 3.3 holds.

9.17 Lemma. X^* is a T_1-space.

Proof. Suppose that $x'^* \in U^*$ implies $x''^* \in U^*$. Let $x'(\alpha|A) \in x'^*$ be of rank 2, and let $x''(\alpha|A) \in x''^*$. Let $U_\alpha = S^3(x'(\alpha),\alpha)$. Then (9.12) $x'^* \in U_\alpha{}^*$; hence $x''^* \in U_\alpha{}^*$; hence $x''(\alpha|A)$ is ultimately in U_α. Hence (9.7) $x'(\alpha|A)$ and $x''(\alpha|A)$ are a Cauchy pair. Hence $x'^* = x''^*$, and (III-8.4) X^* is a T_1-space.

We now let $\{\mathfrak{U}^*(\alpha)|A\}$ be a basis for a struct gX^*. We know (9.16, 9.17) that gX^* by

9.18 $\varphi(x)=\{$**all phalanxes of finite rank converging to** $x\}$.

Further, we clearly have

9.19 ****. $\varphi(x)\in U^*$ **if and only if** $x\in U$.

This implies that $gX \to \varphi(gX)$ is a unimorphism of gX on a substruct of gX^*. From 9.19 we have

9.20 ****. $(S^n(x,\alpha))^* \subset S^n(\varphi(x),\alpha)$.

We use this in

9.21 Lemma. A phalanx, $x(\alpha|A)$, of rank n, belongs to x^* if and only if $\varphi(x(\alpha)|A)$ converges to x^*.

Proof. Let $\varphi(x(\alpha)|A)$ converge to x^*, and let (9.4) $x'(\alpha|A) \in x^*$ be of rank 2. Then (9.14) $x^* \in (S^3(x'(\alpha),\alpha))^*$; hence $\varphi(x(\alpha)|A)$ is ultimately in $(S^3(x'(\alpha),\alpha))^*$; hence (9.17) $x(\alpha|A)$ is ultimately in $S^3(x'(\alpha),\alpha)$. Hence (9.7) $x(\alpha|A)$ and $x'(\alpha|A)$ are a Cauchy pair and $x(\alpha|A) \in x^*$.

Let $x(\alpha|A)\in x^*$; then (9.14 $x^* \in (S^{n+1}(x(\alpha),\alpha))^*$. Therefore $x(\alpha|A)$ is ultimately in $S^{n+1}(x(\alpha),\alpha)$; hence (9.19, 9.20) $\varphi(x(\alpha)|A)$ is ultimately in $(S^{n+1}(x(\alpha),\alpha))^* \subset S^{n+1}(\varphi(x(\alpha)),\alpha)$. Now x^* also belongs to this set, so that $\varphi(x(\alpha)|A)$ is ultimately in $S^{2n+2}(x^*,\alpha)$. Hence (2.11) $\varphi(x(\alpha)|A)$ converges to x^*.

From 8.2, 9.4 and 9.21 we have

9.22 ****. $\overline{\varphi(X)}=X^*$.

We now show that

9.23 Lemma. gX^* is complete.

Proof. Let $x^*(\alpha|A)$ be of rank 2 (in gX^*, with regard to $\{\mathfrak{U}^*(\alpha)|A\}$). Since $\overline{\varphi(X)} = X^*$, we may choose $x'(\alpha)$ so that $\varphi(x'(\alpha)) \in S(x^*(\alpha),\alpha)$. Then (9.8) $\varphi(x'(\alpha)|A)$ is of rank 4, and $x^*(\alpha|A)$ and $\varphi(x'(\alpha)|A)$ are a Cauchy pair. Hence $x'(\alpha|A)$ is of rank 4 and belongs to some x'^*. Hence (9.21, 8.10) $x^*(\alpha|A)$ converges to x'^*, and (9.5) gX^* is complete

We collect all these results (including 8.17) as

9.24 **Theorem. Every struct may be "completed." That is to say, every struct gX has a unimorphic image $\varphi(gX)$ in a complete struct gX^*, where $gX^* = \overline{\varphi(gX)}$. Such a completion is essentially unique, in the sense that any two are unimorphic.**

9.25 Remarks. a) The relative shortness of the corresponding proof for the metric case is due to the use, in that case, of the precise distance, which allows the immediate verification of results corresponding to those which must here be treated separately.

b) The dependence of our completion on a particular basis is convenient, as we shall see below. If the reader wishes a more symmetric treatment, he has only to observe that the process of 9.1 can be applied to the basis for gX consisting of all the large coverings of gX.

Some interesting results follow from the correspondance between a basis for gX and a basis for its completion. We obviously have

9.26 ****. **The completion of a struct with a countable basis is a struct with a countable basis.**

9.27 ****. **The completion of a largely compact struct is a largely compact struct.**

And from these we easily deduce

9.28 ****. **The completion of a metric space is a metric space.**

9.29 ****. **The completion of a largely compact struct is a compact struct.**

9.30 ****. **A struct is largely compact if and only if its completion is compact.**

There are many interesting and unsolved problems concerning the connection between a struct and its completion. Is, for instance, the completion of an a-struct an a-struct?

10. Historical Remarks. The notion of "uniformity" has been investigated recently by several mathematicians (Cohen 1937, Cohen 1939, Graves 1937, Weil 1936, Weil 1937). The most complete treatment is that of Weil. This is the only one we will discuss, except to observe that Cohen uses a local uniformity rather than a uniformity in the large.

To discuss the relation between Weil's methods and our own, we use the familiar terminology of the metric case. Weil's central entities are the e-spheres of center x ; ours are the coverings by sets of diameter less than e . Weil considers the topology derived from the metric; we consider those metrics which agree with the topology. In both cases we must consider entities outside the space itself. Weil uses "neighborhoods" of the diagonal of the product of the space by itself; we use open coverings of the space.

The two treatments, Weil's and our own, come necessarily to the same results. The majority of the results considered here are originally due to Weil, although the methods of proof and the fundamental notions are often different.

I prefer the use of coverings, for they are important in other branches of topology. Coverings and convergence are two concepts in the main stream of general topology, and therefore are especially suited to the discussion of uniformity. I hope, in the near future, to apply these methods to a satisfactory dimension theory of structs, using the ideas of Čech.

In conclusion, we should observe that all the domains of the analyst and the algebraist which have "adequate" topologies are structs.

Chapter VII

FUNCTION-SPACES

1. Introduction.
2. Spaces.
3. Structs.
4. Compactness.
5. Parallelotopes and Embedding.
6. Characteristic Functions and Set Spaces.

1. Introduction. In this chapter we consider certain spaces of functions which have topological significance. One important example of such a function space is the topological (or Cartesian) product of several spaces.

In §2 we examine these function spaces as spaces, in §3 we examine them as structs. In §4 we consider their compactness. In §5 we consider parallelotopes and prove Tychonoff's embedding theorem as generalized to largely compact structs. In §6 we consider the spaces made up of characteristic functions and of subsets.

2. Spaces. We shall consider only two sorts of function-spaces, one in which convergence means pointwise convergence, and one in which convergence means uniform convergence.

If, for each $a \in A$, X^a is a set, then (I-5) $P\{X^a|A\}$ is a set whose elements are all the functions $x(a|A)$ for which $x^a = x(a) \in X^a$ for all a. We call $x^a = x(a)$ the ath coordinate of x. The function p^a defined by $p^a(x) = x$ is called the projection on the ath coordinate.

If each X^a is a space, then we proceed to make $P\{X^a|A\}$ a space. Fix $x_1 = \{x_1^a|A\}$; for some a choose a nbd N^a of x_1^a in X^a, and consider $N = \{x|x^a \in N^a\}$. We call this set a hyperslice with base N^a. If we let N^a run thru all a sub-basic set of nbds of x_1^a, and then let a run thru A, we obtain a collection of sets which we define to be a sub-basic set of nbds of x_1, By the methods of III-3 we may easily see that

2.1 **. If $x(b|\mathfrak{B})$ is a mapping of a directed system into $X = P\{X^a|A\}$, then $x(b|\mathfrak{B})$ converges (in X) to x if and only if $x^a(b|\mathfrak{B})$ converges (in X^a) to x^a for each a.**

2.2 **. If $f=f(y|Y)$ is a function from the space Y to $X=P\{X^a|A\}$, then it is continuous if and only if the functions $f^a=f^a(y|Y)=p^a(f(y)|Y)$ from Y to X^a are continuous for each a.**

We call $\{X^a|A\}$ the topological or pointwise product of the X^a.

If all the X^a are T-spaces, then $\{X^a|A\}$ is also a T-space. A sub-basis for its open sets is formed by the hyper-slices with open bases.

If all $X^a = X$, we write $(X^{|A|})_p$ for $\{X^a|A\}$. If $|A| = |B|$, then clearly $(X^{|A|})_p$ is homeomorphic to $(X^{|B|})_p$. By writing $(X^{|A|})_p$ we imply that we use A as the set of indices in forming the product.

In simple cases this product is very familiar; if I is a segment, then $(I^3)_p$ is a cube.

We note that a product of discrete spaces is discrete if and only if there are a finite number of factors.

We now consider a different type of function space, one where convergence means uniform convergence. If gX is a struct and A is a set (discrete space), then the points of $(gX^{|A|})_u$ are those of $(X^{|A|})_p = P\{X^a|A\}$, where $X^a = X$. That is, the points of $(gX^{|A|})_u$ are all the functions from A into gX (if we consider A as a discrete struct, then these functions are all the uniformly continuous functions from the discrete struct into gX). If $\varphi, \psi \in (gX^{|A|})_u$, and if $\mathfrak{U}(b)$ is a large covering in gX, then $\psi \in U_b^A(\varphi)$ if, for all a, $\psi(a) \in S(\varphi(a), b)$. If $\{\mathfrak{U}(b)|B\}$ is a basis for gX, then we define $\{U_b^A(\varphi)|b \in B, \varphi \in (gX^{|A|})_u\}$ as a basis for the open sets of $(gX^{|A|})_u$. It is clear that two such bases obtained from different bases of gX are equivalent. We asserted that convergence in $(gX^{|A|})_u$ means uniform convergence. This follows from

2.3 **. If $y(d|\mathfrak{D})$ is a mapping of a directed system into $Y=(gX^{|A|})_u$, where $y(d)=x_d(a|A)$, and $y=y(a|A)$, then $y(d|\mathfrak{D})$ converges to y if and only if, for each large covering $\mathfrak{U}$ of gX there is a d' such that $d>d'$ implies $x_d(a) \in S(x(a),\mathfrak{U})$ for all a.**

2.4 **. A function φ from Z to $Y=(gX^{|A|})_u$ is continuous if and only if it is equicontinuous in a. That is, if and only if, setting $\varphi(z)=x_z(a|A)$, for each $z' \in Z$ and each large covering $\mathfrak{U}$ of gX there exists an open set $V \ni z'$, such that $z \in V$ implies $x_z(a) \in S(x_{z'}(a),\mathfrak{U})$ for all a.**

If A is finite, and if X is a normal T_1-space, then we see that $(fX^{|A|})_u$, $(aX^{|A|})_u$ and $(X^{|A|})_p$ are homeomorphic.

If hY is another struct, then gX^{hY} is defined as the subspace of $(gX^{|Y|})_u$ made up of uniformly continuous functions.

2.5 Theorem. gX^{hY} is closed in $(gX^{|Y|})_u$. That is, a uniform limit of uniformly continuous functions is uniformly continuous.

Proof. Let $\{\mathfrak{U}(a)|A\}$ be a basis for gX and let $\{\mathfrak{V}(b)|B\}$ be a basis for hY. Let $z(\delta|D)$ converge to z, where $z(\Delta) \subset gX^{hY}$. Let $z(\delta) = z\delta(y|Y)$ and $z = z(y|Y)$. Let a be fixed, and choose a' so that (VI-2.11) $S^3(H,a') \subset S(H,a)$ for all $H \subset X$. Then (2.3) we may fix δ' so that $z_{\delta'}(y) \in S(z(y),a')$ for all y. Since $z_{\delta'} = z(\delta')$ is uniformly continuous, there exists (VI-5.3) a b such that $y' \in S(y'',b)$ implies $z_{\delta'}(y') \in S(z_{\delta'}(y''),a')$. Combining this with the results above, we see that $y' \in S(y'',b)$ implies $z(y') \in S(z(y''),a)$. Hence we have shown that z is uniformly continuous.

For aX^{hY} we write simply X^{hY}, for gX^{aY} we write gX^Y, and for aX^{aY} we write X^Y. Since every continuous function on an a-struct is uniformly continuous, gX^Y consists of all continuous functions on Y to X, where convergence means uniform convergence in gX. If X and Y are both compact, then $gX = aX$ and $hY = aY$ so that $gX^{hY} = X^Y$.

3. Structs. We wish to attach a uniformity to $P\{gX^a|A\}$. If $\mathfrak{U}^a$ is large in gX^a, we naturally consider the covering $\mathfrak{U}$ whose open sets are the hyperslices based on the sets of $\mathfrak{U}^a$. The class of coverings obtainable in this way is defined to be a sub-basis for the large coverings of $P\{gX^a|A\}$. We see immediately (VI-3.4) that this uniformity agrees with the topology of $P\{gX^a|A\}$ which is that of $\{X^a|A\}$. Hence we have

3.1 **. If each gX^a is a struct, then $P\{gX^a|A\}$ is a struct.**

We may easily see that

3.2 **. A phalanx in $P\{gX^a|A\}$ is Cauchy if and only if each of its ath projections is a Cauchy phalanx in the corresponding gX^a.**

Hence (2.1) we have

3.3 **. gX is complete if and only if each gX^a is complete.**

We see directly, that

3.4 **. gX is largely compact if and only if each gX^a is largely compact.**

Combining these results (VI-8.14), we obtain a special case of an important result of the next § .

3.5 **. gX is compact if and only if each gX^a is compact.**

If all $gX^a = gX$, we write $(gX^{|A|})_p$ for $P\{gX^a|A\}$.

We now return to $(gX^{|A|})_u$ and the notation of the last § . We define $\mathfrak{U}^A(b)$ by $\mathfrak{U}^A(b) = \{U_b^A(\varphi)|\varphi \in (gX^{|A|})_u\}$. If $\{\mathfrak{U}(b)|B\}$ is a basis for gX , then we define $\{\mathfrak{U}^A(b)|B\}$ to be a basis for $(gX^{|A|})_u$. It is easy to see that any two such bases are equivalent and that the uniformity so defined agrees with the topology. Hence $(gX^{|A|})_u$ is a struct.

We have

3.6 Theorem. If gX is complete, then $(gX^{|A|})_u$ is complete.

Proof. Let $\{\mathfrak{U}(\beta)|B\}$ be a basis of gX satisfying VI-9.1. Then $\{\mathfrak{U}^A(\beta)|B\}$ is a similar basis for $(gX^{|A|})_u$. If $z(\gamma|C)$, where $z(\gamma) = \{x^a(\gamma)|A\}$, is of rank 2 in $(gX^{|A|})_u$, then (VI-9.2) $x^a(\gamma') \in S^2(x^a(\gamma''),\gamma)$ whenever $\gamma' \supset \gamma$ and $\gamma'' \supset \gamma$. Therefore, for each a , $x^a(\gamma|C)$ is of rank 2 in the complete struct gX and hence has a limit x^a . Since $x^a(\gamma|C)$ is ultimately in $S^2(x^a(\gamma),\gamma)$, and ultimately in $S(x^a,\gamma)$, these sets meet, and $S^2(x^a(\gamma),\gamma) \subset S^6(x^a,\gamma)$. Therefore, for each a , $\gamma' \supset$ implies $x^a(\gamma') \in S^6(x^a,\gamma)$. Hence (2.3, VI-2.11) $z(\gamma|C)$ converges to $z = \{x^a|A\}$. Hence (VI-8.14) $(gX^{|A|})_u$ is complete.

We regard gX^{hY} as a substruct of $(gX^{|Y|})_u$. From 2.5 and 3.6 we have (VI-8.17)

3.7 **. If gX is complete, then gX^{hY} is complete.**

The converses of 3.6 and 3.7 are trivial.

We close this section by asserting certain unimorphisms. It is easy to show, in each case, that the obvious correspondence is a unimorphism.

3.8 **. We have.**

a) If gX is unimorphic to hY and |A| |B|, then $(gX^{|A|})_p$ is unimorphic to $(hY^{|B|})_p$, and $(gX^{|A|})_u$ is unimorphic to $(hY^{|B|})_u$.

b) If gX is unimorphic to hY and jV is unimorphic to kZ, then gX^{jV} is unimorphic to hY^{kZ}.

4. Compactness. We begin with

4.1 Theorem. $X = P\{X^a | A\}$ **is compact if and only if each** X^a **is compact.**

Proof. Suppose each X^a compact. Let $x(\gamma|C)$ be an ultraphalanx in X. Since each projection is single-valued, each $x^a(\gamma|C)$ is (IV-3.1) an ultraphalanx, which converges (IV-4.6), say to x^a. Then (2.1) $x(\gamma|C)$ converges to $x = \{x^a|A\}$. Hence (IV-4.6) X is compact. The converse is left to the reader.

The analogous result about hX^{gY} is false. Consider I^I, where I is the closed interval $[0,1]$. Define $x(\alpha) = f_\alpha(t|I)$ by

$$f_\alpha(t) = \begin{cases} 0, & t = 0, \\ |\alpha|^{-1}, & t = 1-|\alpha|^{-1} \\ 1, & t = 1. \end{cases}$$ (where $|\alpha|$ is the cardinal number of the set α.)

and making $f_\alpha(t|I)$ linear elsewhere. In $(I^{|I|})_p$, the phalanx $x(\alpha|A)$ has the unique limit $x' = f'(t|I)$, where

$$f'(t) = \begin{cases} 0, & t < 1, \\ 1, & t = 1. \end{cases}$$

Now in $(I^{|I|})_u$, which has a finer topology than $(I^{|I|})_p$, this phalanx fails to converge to this limit and fails to have x' as a cluster point. Since x' is the only cluster point of this phalanx in the coarser topology, it follows that the phalanx has no cluster point in the finer topology. This is also true in I^I, so that this struct also fails to be compact.

5. Parallotopes and Embedding. A parallelotope is a pointwise power $(I^{|A|})_p$ of the closed interval $[0,1]$. It is well known (Heine-Borel theorem and IV-4.13) that I is compact; hence each parallelotope is compact.

We shall need certain bases and sub-bases for a parallelotope. $U_e^\alpha(x_o)$ consists of all $x \in (I^{|A|})_p$ for which $|x^a - x_o^a| < e$ for $a \in \alpha$. (Here, of course, e is a positive real number.) We have

5.1 **.** $\{U_e^\alpha(x) | \alpha \in A,\ e > 0\}$ **is a nbd basis at** x.
$\{U_e^a(x) | a \in A,\ e > 0\}$ **is a nbd sub-basis at** x.

$\mathfrak{U}_e^\alpha$ **consists of all** $U_e^\alpha(x)$; **we abbreviate** $S(x, \mathfrak{U}_e^\alpha)$ **as** $S(x, \alpha, e)$.

5.2 **.** $\{\mathfrak{U}_e^\alpha | \alpha \in A,\ e > 0\}$ **is a basis for the struct** $(I^{|A|})_p$.
$\{\mathfrak{U}_e^a | a \in A,\ e > 0\}$ **is a sub-basis for the struct** $(I^{|A|})_p$.

We have

5.3 Theorem. Every compact struct is unimorphic to a substruct of a parallelotope.

Proof. If gX is compact, then (VI-4) X is a normal T_1-space. If $\{U_b | B\}$ is a basis for gX, and if $\{\mathfrak{U}_a | A\}$ are the basic binary coverings belonging to this basis (V-4.7), then the real-valued continuous functions (VI-9.3) $\varphi^a(x|X)$, where

$$\varphi^a(x) = \begin{cases} 0, & x \in \overline{U_{b''}} , \\ 1, & x \in X - U_{b'} , \end{cases}$$

and where $\mathfrak{U}_a = \{U_{b'}, X-\overline{U_{b''}}\}$, map X into $(I^{|A|})_p$. Since each φ^a is continuous, it follows (2.2) that $\varphi = \{\varphi^a | A\}$ is continuous and hence (VI-5.5) uniformly continuous.

Now from V-4.7 we see that φ^{-1} is single-valued. Let $x \in U \subset X$, where U is open, then (VI-4.7), there is a basic binary covering $\{U', X-\overline{U''}\}$ where $x \in U'' \subset U' \subset U$, hence $\varphi^a(x) = 0$, and if $\varphi^a(x) < 1$, then $x \in U' \subset U$. Since $\varphi(X) \cap \{x | \varphi^a(x) < 1\} \subset (\varphi^{-1})^{-1}(U)$ is a nbd of $\varphi(x)$ in $\varphi(X)$, we see that φ^{-1} is continuous. As a continuous image (IV-5.1) of X, $\varphi(X)$ is compact, hence (VI-5.5) φ^{-1} is uniformly continuous.

Since every largely compact struct is unimorphic to a substruct of a compact struct (its completion), we have

5.4 **. Every largely compact struct is unimorphic to a substruct of a parallelotope.**

Conversely, every struct unimorphic to a substruct of a parallelotope is largely compact.

Since an f-struct can be erected on any completely regular space, we have

5.5 **. (Tychonoff's Theorem) Every completely regular space is homeomorphic to a subspace of a parallelotope.**

Conversely, every space homeomorphic to a subspace of a parallelotope is completely regular.

6. Characteristic Functions and Set Spaces. Consider $(T^A)_p$, where T is the discrete space consisting of the two points 0 and 1. The points of $(T^{|A|})_p$ are functions on A taking the values 0 and 1. Such a function is usually called the characteristic function of the set on which it takes the value 1. This defines a one-to-one correspondence between the points of $(T^{|A|})_p$ and those of $2^{|A|}$. We are naturally led to introduce the topology in $2^{|A|}$ which makes this a homeomorphism.

If we do this, then we have

6.1 **. A mapping $C(b|\mathfrak{B})$ of a directed system into $2^{|A|}$ converges in $2^{|A|}$ to $C \subset A$ if and only if, for all $x \in C$, $C(b|\mathfrak{B})$ ultimately contains x, and for all $x \notin C$, $C(b|\mathfrak{B})$ ultimately fails to contain x. That is, if and only if**

6.2 $C = \cup\{\cap\{C(b)|b>b'\}|b' \in \mathfrak{B}\} = \cap\{\cup\{C(b)|b>b'\}|b' \in \mathfrak{B}\}$.

We may easily recognize 6.2 as a natural generalization of a well-known condition for the convergence of sequences of sets.

Since T was a compact struct, we have

6.3 **. $2^{|A|}$ is a compact struct.**

We need the following result later,

6.4 **. If $C(n+1) \supset C(n)$ for all n, then $C(n|N)$ converges to $\cup\{C(n)|N\}$.**

We may regard $2^{|A|}$ as a ring, defining the "sum" of B and C to be $(B \cup C)-(B \cap C)$, and the "product" of B and C to be B C. It is easy to see that these operations are continuous in the topology we have introduced. Thus $2^{|A|}$ is a topological ring.

Chapter VIII

EXAMPLES

1. Introduction.
2. Some Non-normal Spaces.
3. Some Simple Structs.
4. Some Mildly Complex Structs.

1. Introduction. We collect here a few examples of interest from a rather general point of view.

1.1 In this chapter we remove the restriction that γ is a finite subset of C. We use γ for any subset of C, and Γ for any collection of γ's.

2. Some non-normal spaces. We use a simple modification of an idea of Čech to exhibit some examples of spaces which are completely regular but not normal.

Let C be any uncountable set (e.g., the real numbers), and consider $\Gamma = 2^{|C|}$, with the topology discussed in VII-6. Let $\Gamma_2 = \{\gamma | \gamma$ is countable$\}$; then Γ_2 is a subspace of Γ_1 . We now consider the product $P(\Gamma_1, \Gamma_2)$. We shall write its elements as ordered pairs. We consider two subsets of the product,

$$D = \{(\gamma,\gamma) | \gamma \in \Gamma_2\}$$
$$E = \{(C,\gamma) | \gamma \in \Gamma_2\} \quad , \ (C \text{ does belong to } \Gamma_1 \,!).$$

I say that these sets are both closed. D is closed, for if $(\gamma(\alpha|A),\gamma(\alpha|A))$ converges to (γ',γ'') it is clear (VII-6.2) that $\gamma' = \gamma''$. E is closed, for the point C is closed in Γ_1 . Let U and V be open sets with $D \subset U$ and $E \subset V$. Let $\gamma(1) \in \Gamma_2$. Let $\varphi(\gamma) = \gamma$, then $\varphi(\gamma|C)$ converges to C , and hence $(\varphi(\gamma|C),\gamma(1))$ converges to $(C,\gamma(1)) \in V$. Hence there is an $\gamma(2) \supset \gamma(1)$ for which $(\gamma(2),\gamma(1)) \in V$.

We repeat this argument to obtain a simple sequence $\{\gamma(n) | N\}$, so that $(\gamma(n+1),\gamma(n)) \in V$, and $\gamma(n+1) \supset \gamma(n)$. Let $\gamma' = \cup\{\gamma(n) | N\}$; then (VII-6.4) both $\gamma(n)$ and $\gamma(n+1)$ converge to γ' as $n \to \infty$. Hence $(\gamma(n+1),\gamma(n))$ converges to (γ',γ') as $n \to \infty$. But $(\gamma',\gamma') \in D \subset U$, hence, for some n , $(\gamma(n+1),\gamma(n)) \in U$;

hence $U \cap V \neq \emptyset$. Thus $P(\Gamma_1, \Gamma_2)$ is not normal.

Now (VII-3, VII-6) $\Gamma_1 = 2^{|C|}$ is a struct in a natural way, and so is $P(\Gamma_1, \Gamma_2)$; hence the product is a completely regular space. So we have

2.1 $P(\Gamma_1, \Gamma_2)$ is completely regular but not normal.

If we regard Γ_1 as a ring, then Γ_2 is a sub-ring, and we make $P(\Gamma_1, \Gamma_2)$ a ring in the natural way. Hence

2.2 **. There exist non-normal topological rings.**

By replacing T ($T^{|C|}$ was homeomorphic to $2^{|C|}$ (VIII-6)) by, 1) the group of the real numbers modulo 1 (the circle group), or 2) the group of all real numbers, it may be shown that

2.3 **. There exist connected, completable, non-normal topological groups.**

2.4 **. There exist non-normal linear topological spaces.**

3. <u>Some simple structs</u>. We assume that the topology of the closed interval of real numbers [0,1] is known. We shall be interested in two of its simpler subspaces. Let X consist of the points 1/n, where n is a positive integer. Then X is discrete and fully normal; hence every finite covering of X is open and normal. Let Y consist of the points of X and the point 0 . Each open covering of Y contains an open set containing 0 , and hence containing all points 1/n for n greater than some n' .

Suppose that gY is a struct, and that gX is a sub-struct of gY . Then each large covering of gX contains a set containing all the points 1/n for n greater than some n' . fX lacks this property; hence fX is a substruct of no gY .

4. <u>Some mildly complex structs</u>. We start with any uncountable set C , and the family Γ_e of its countable subsets (1.1). Zorn's Lemma asserts that there exists a maximal linearly ordered (by $\supset$, of course) subfamily of Γ_e . Let one such be Γ . We shall need

4.1 If $\gamma_n \in \Gamma$ for each n, then $\cup\{\gamma_n | N\}$.

To prove this, let $\gamma' = \cup\{\gamma_n | N\}$, and let $\gamma \in \Gamma$. If, for some n , $\gamma \subset \gamma_n$, then $\gamma \subset \gamma'$. If, for all n , $\gamma_n \subset \gamma$, then $\gamma' \subset \gamma$. We see that $\Gamma' = \Gamma \cup \{\gamma'\}$ is linearly ordered. Since Γ was maximal, we have $\Gamma' = \Gamma$ and $\gamma' \in \Gamma$. Thus 4.1 is proved. Similarly, we have

4.2 If $\Gamma_1 \subset \Gamma$ and if $\cup\{\gamma' | \gamma' \in \Gamma_1\}$ is a countable set, then this set belongs to Γ .

We now introduce what might be called a "lower topology" into Γ. If $\gamma_1 \in \Gamma$ and $\gamma \neq \emptyset$, then a basis for the nbds of γ_1 is made up of the sets $N_\gamma(\gamma_1)$, where $\gamma' \subset \gamma_1$ and

$$N_{\gamma'}(\gamma_1) = \{\gamma'' \mid \gamma' \subset \gamma \subset \gamma_1 \quad , \quad \gamma \neq \gamma'\} .$$

A basis for the nbds of $\emptyset$ is $\{\{\emptyset\}\}$. This topology clearly makes Γ a space, in which we have

4.3 If $\Gamma_1 \subset \Gamma$, and if $\cup\{\gamma \mid \gamma \in \Gamma_1\}$ belongs to Γ, then it belongs to $\overline{\Gamma}_1$.

4.4 If $\gamma_{n+1} \supset \gamma_n$, then $\{\gamma_n \mid N\}$ converges to $\cup\{\gamma_n \mid N\}$.

We now show that

4.5 Γ is completely normal.

Let Γ_1 and Γ_2 be subsets of Γ for which $\overline{\Gamma}_1 \cap \Gamma_2 = \emptyset = \Gamma_1 \cap \overline{\Gamma}_2$. If $\gamma_1 \in \Gamma_1$, then we consider $\gamma' = \{\gamma'' \mid \gamma'' \in \Gamma_2\}$, $\gamma'' \subset \gamma_1$, we see (4.2, 4.3) that $\gamma' \in \overline{\Gamma}_2$; hence $\gamma' \neq \gamma_1$ and $N_{\gamma'}(\gamma_1)$ is an open set which does not meet Γ_2. We proceed in the same way for each $\gamma_1 \in \Gamma_1$ and in the corresponding way for each $\gamma_2 \in \Gamma_2$. Then $\Gamma_3 = \cup\{N_{\gamma'}(\gamma_1) \mid \gamma_1 \in \Gamma_1\}$ and $\Gamma_4 = \cup\{N_{\gamma''}(\gamma_2) \mid \gamma_2 \in \Gamma_2\}$ are open sets, such that $\Gamma_1 \subset \Gamma_3$, $\Gamma_2 \subset \Gamma_4$, and $\Gamma_1 \cap \Gamma_4 = \emptyset = \Gamma_2 \cap \Gamma_3$. Now from the linearity of Γ and the construction of Γ_3 and Γ_4 we see that Γ_3 and Γ_4 are disjoint. Hence Γ is completely normal.

For each $\gamma \in \Gamma$, we define $S_\gamma = \{\gamma'' \mid \gamma' \supset \gamma, \gamma' \neq \gamma, \gamma' \in \Gamma\}$. Then we have

4.6 If $\mathfrak{U}$ is an open covering of Γ, then some S_γ is included in some $V \in \mathfrak{U}^\Delta$.

If not, for each γ and some $\gamma_a(\gamma) \supset \gamma$, $\gamma_a(\gamma) \in S(\gamma, \mathfrak{U})$. Let γ_1 be arbitrary, and put $\gamma_{n+1} = \gamma_a(\gamma_n)$. Let $\gamma' = \cup\{\gamma_n \mid N\}$, which (4.1) belongs to Γ. Let $\gamma' \in U \in \mathfrak{U}$. Since (4.4) $\{\gamma_n \mid N\}$ converges to γ', there is an n', such that $\gamma_n \in U$, for $n > n'$. Hence $\gamma_{n'+1} = \gamma_a(\gamma_{n'}) \in S(\gamma_{n'}, \mathfrak{U})$; which is a contradiction. Thus 4.6 is proved, and we have at once

4.7 Every normal covering of Γ contains a set including an S_γ.

We now see that

4.8 $a\Gamma$ is not complete.

Let $D = \Gamma$, then for $\delta \in \Delta$ (We still use δ for <u>finite</u> subsets of D) let $\gamma(\delta) = \cup\{\gamma \mid \gamma = d \in \delta\}$. I say that $\gamma(\delta \mid D)$ is a Cauchy phalanx. For clearly $\gamma(\delta \mid D)$ decides for each S_γ, and hence (4.7) $\gamma(\delta \mid D)$ is Cauchy. But, for the same reason, $\gamma(\delta|D)$ cannot converge to any element of Γ, since $X-S_\gamma$ is a nbd of γ.

This example is due, in a slightly different form, to Dieudonné 1939.

Chapter IX

DISCUSSION

1. What is Topology?
2. The Role of Denumerability.
3. Which Separation Axioms are Important?
4. No Transfinite Numbers Wanted.
5. The Subsequence and Its Generalizations.
6. Phalanxes vs. Filters.

1. What is topology? This heading is inaccurate; the meaningful question and the one we discuss is "What should we mean by topology?" . We must begin by making it clear that we do not (in this discussion) include algebraic or combinatorial topology. We return to this point later.

Classically, topology was sometimes defined as "the geometry whose group is the group of all bicontinuous transformations." I feel that this definition is both too wide and too restricted: it uses the word "geometry," and it insists on invariance under all bicontinuous transformations. Thus it would include combinatorial topology and exclude any discussion of uniformity (as a part of topology).

Topology should be an analog of modern algebra. Modern algebra is concerned with suitably restricted finite operations and relations. Topology should be concerned with suitably restricted infinite operations and relations. (In both cases, we may expect "suitably restricted" to have a force diminishing with increasing time.) Prominent examples of such infinite operations are convergence and closure. While these can be regarded as finite operations, this point of view usually serves only to hide their true nature, as this nature is revealed in their applications. From this point of view, uniform convergence, uniform continuity, and their related ideas, are a very proper part of topology.

When we include uniformity in topology, we must be careful to exclude metric notions. The notion of pairs of equidistant points is surely not a topological one; it belongs to a discipline of more detail and less generality.

Since topology is abstract and general (even for mathematics), we may expect its ideas and results to be useful tools in many distinct branches of mathematics. The need, in several of these branches, notably analysis and topological algebra, for a suitable treatment of uniformity proves an additional reason for making uniformity a part of topology.

According to this view, algebraic topology and topological algebra are the results of combining algebra and topology. In the first, algebra is applied to topological objects; in the second, topology is applied to algebraic objects.

2. The role of denumerability. The unusual, and seemingly unnecessary, position of denumerability in topology has caused comment (Weil 1937). There seems to be no *a priori* reason why this should be so. Yet metrization (the possibility of a struct with a countable basis, VI-6.1), a countable basis, the first countability axiom, and other references to denumerability appear in the hypotheses of many topological theorems. In some of these theorems (a regular space with a countable basis is normal, a complete metric space is never of the first category in itself, $\cdots$) the hypothesis of denumerability cannot be replaced by an essentially more general one without destroying the theorem.

After considering the role of directed systems in these arguments, one finds that denumerability is essential only to ensure the equivalence of sequence and phalanx, the conjunction of monotonicity and finiteness. A set is countable if and only if it can be "reached" ("swelled to") by a monotone family of its finite subsets. The countable is important because it is so nearly finite.

There is no other infinite cardinal showing similar special properties in topology. This has been experimentally true in the past and can confidently be expected to hold in future.

3. Which separation axioms are important? What classes of objects are important for topological study? We have examples of spaces that are T- but not T_0-, T_0- but not T_1-, T_1- but not Hausdorff, Hausdorff but not regular, regular but not completely regular, completely regular but not normal, normal but nor completely normal, completely normal but not perfectly normal, perfectly normal but not metrizable, and so on. Are all these categories important? Shall we divide our energies among them all? To me it seems that only a few are important; this can be verified only after more experience; but I feel that one can now see dimly which are the ones that matter.

The most general class of object that I would suggest for topological study at present is what is defined in III-3.16 as a space. We require, for example, of the closure operator only that it commute with union and preserve the empty set. I am not prepared to assert that this is ultimate generality, but rather that this is a natural class of sufficient generality for present purposes.

It may be true that T-spaces form an important class; I am inclined to doubt this.

The next important class is that of completely regular spaces, which contains all algebraic objects with an adequate topology. The importance of this class comes from the fact that on them we may erect structs. Since we may build two topologically invariant structs (which coincide in the compact case) on every completely regular space, it is not unreasonable to say that if a space is completely regular we should usually regard it as a struct.

I do not think that normality is important. The introduction (V-3) of the notion of a normal covering should allow many of the important results about normal spaces to be extended to completely regular spaces. And on the other hand, there exist examples (VIII-2) of various kinds of algebraic objects which fail to be normal. I believe that no algebraic condition (not involving or implying finiteness or countability) can ensure normality.

Full normality remains to be investigated; it may be that it has importance in dimension theory and related topics.

Metrizability is of uncertain importance; a large part of its present position may come from its implication of full normality.

We have yet to discuss the requirements that a T-space be T_0-, T_1-, Hausdorff, or regular. I allot to these a very minor role. They are convenient if another hypothesis allows one to pass from them to complete regularity (as in the theorem that every T_1-topological group is completely regular). They may perhaps be useful in certain special places. To one who has seen "Hausdorff space" so many, many times this may seem rather harsh treatment; but if he examines a few examples I believe that he will eventually agree with this position.

I close this discussion by pointing out the need for a shorter and more suitable term than "completely regular space," and proposing the term "Tychonoff space."

4. No transfinite numbers wanted. I believe that transfinite numbers, particularly ordinals, have a proper place only in descriptive theories, such as: the successive derivatives of a

set, the Borel classes of sets and the Baire classes of functions, and some of the less pleasing parts of the theory of directed systems. We have succeeded in eliminating all infinite ordinals from the treatment of the subjects dealt with here. The only transfinite cardinal to make an essential appearance is $\aleph_0$, and we have already seen that denumerability does play an important and distinctive role.

Zorn's Lemma serves to eliminate arguments by transfinite induction. The other stronghold of the transfinite in topology has been the construction of counterexamples. In VIII-2 and VIII-4 we see two ways of adapting examples, originally constructed with transfinite ordinals, to examples constructed with sets.

5. The subsequence and its generalizations. The stumbling block in the effort to generalize satisfactorily the convergence of sequences was the subsequence. We may define a subsequence in several ways, and, if we try to generalize the wrong one, the results may be unfortunate.

We may look on a subsequence as a cofinal part of a sequence, as a part that goes beyond any given point. If we realize that linear order is inessential, then we are led to the ideas of Moore and Smith and to a satisfactory theory of convergence.

We may look on the statement "every sequence contains a convergent subsequence" as a way of stating that there are one or more points from which the sequence "cannot tear itself away." If we adopt this view, we are led to cluster points and a satisfactory theory of compactness.

We may look on the statement "every sequence contains a convergent subsequence" as being in the form we wish to generalize. If we do, we find that it will not generalize satisfactorily.

We may look on a sequence as a "point-set" having the same cardinal number as the "point-set" corresponding to the whole sequence. If we do, we are led to the notion of "complete limit-point" and the complications of infinite cardinal arithmetic. This leads to a complicated and non-perspicuous theory of compactness.

This last point of view was encouraged by two details. First, the analyst's habit of saying "for infinitely many n," when he meant "for a subsequence of n." As long as one is solely concerned with sequences, this is a simple and suitable method of expression. From a general point of view, however, this expression emphasizes the wrong property. Second, in talking about a sequence, in topology but not in analysis, there was a tendency to neglect the ordering; to regard a sequence as a "point-set" rather than as a function defined on the positive integers. This attitude is exemplified by the careful (sometimes!) distinction between a "Punktfolge" and a "Zahlfolge."

It is my firm conviction that this last point of view is not a good one, as judged by its results; for I feel that cardinal arithmetic of the complicated sort ("regular alephs," "accessible alephs," etc.) should be kept as far from general topology as possible. This is the ordinal part of the theory of cardinal numbers, and is essentially descriptive. It is not the task of general topology to describe objects in terms of ordinal numbers.

6. Phalanxes vs. Filters. The notion of a filter, introduced by H. Cartan in 1937, has been used by N. Bourbaki and his collaborators as a generalization of a sequence. That is, they consider the "convergence" of filters.

Given a sequence, they consider the filter of all sets which contain the image of a "residue" of the integers. The filter converges to a point if it contains every nbd of the point. Thus for filters derived from sequences we have the classical convergence condition.

This notion has the disadvantage that, in the case of sequential convergence, it concerns itself with the unintuitive family of sets, rather than the rather intuitive sequence. Thus, to obtain generality, we must abandon the intuitively satisfactory treatment of the sequential case. It may seem unfortunate, to one interested in generalities, that the sequence has a special place in topology. It seems, however, to be true. (We may always replace the sequence by a phalanx on a countable base, but in a few places this seems artificial).

Phalanxes, on the other hand, form a part of a theory of convergence (of functions on directed systems) which includes sequences. We obtain generality without discarding the intuitive treatment of special cases.

In closing, it should be noted that the idea of an ultraphalanx stems from the idea of an ultrafilter, and that they are essentially equivalent. The equivalence is between a single ultrafilter and a class of ultraphalanxes. Ultrafilters have the advantage of uniqueness. Ultraphalanxes have the advantages of occasional simplicity and inclusion in a simple theory of convergence.

Bibliography

Alexandroff and Urysohn 1933.
P. Alexandroff and P. Urysohn, "Sur les espaces topologiques compacts." Acad. Sci. Cracovie Bull. Int. (also known as "Bull. Int. Acad. Polonaise Sci. et Lett.) 1923, pp. 5-8 (1924).

Banach 1932.
S. Banach, "Théorie des opérations linéaires." Monografje Matematyczne 1, (Warsaw).

Birkhoff 1937.
Garrett Birkhoff, "Moore-Smith convergence in general topology." Annals of Math. 38, pp. 39-56.

Birkhoff 1939.
Garrett Birkhoff, "An ergodic theorem for general semi-groups" Proc. Nat. Acad. Sci. 25, pp. 625-627.

Cartan 1937.
H. Cartan, "Théorie des filtres" and "Filtres et ultrafiltres." Comptes Rendus (Paris) 205, pp. 595-598 and 777-779.

Cohen 1937.
L. W. Cohen, "Uniformity properties in a topological space satisfying the first denumerability postulate." Duke Math. J. 3, pp. 610-615.

Cohen 1939.
L. W. Cohen, "On imbedding a space in a complete space." Duke Math. J. 5, pp. 174-183.

Dieudonné 1939.
J. Dieudonné, "Un example d'espace normal non susceptible d'une structure uniforme d'espace complet." Comptes Rendus (Paris) 209, pp. 145-147.

Fréchet 1906.
M. Fréchet, "Sur quelques points du calcul fonctionnel." Rendiconti Palermo 22, pp. 1-74.

Fréchet 1921.
M. Fréchet, "Sur les ensembles abstraits." Ann. Ecole Norm. 38, pp. 341-385.

Fréceht 1928.
M. Fréchet, "Les espaces abstraits." Paris, Gauthier-Villars.

Frink 1937.
A. H. Frink, "Distance-functions and the metrization problem." Bull. Amer. Math. Soc. 43, pp. 133-142.

Graves 1937.
L. M. Graves, "On the completing of a Hausdorff space." Annals of Math. 38, pp. 61-64.

Moore 1915.
E. H. Moore, "Definition of limit in general integral analysis." Proc. Nat. Acad. Sci. (Washington) 1, pp. 628-632.

Moore and Smith 1922.
E. H. Moore and H. L. Smith, "A general theory of limits." Amer. Jour. of Math. 44, pp. 102-121.

Weil 1936.
A. Weil, "Les recouvrements des espaces topologiques: espaces complets, espaces bicompacts." Comptes Rendus (Paris) 202, pp. 1002-1005.

Weil 1937.
A. Weil, "Sur les espaces à structure uniforme et sur la topologie générale." Actualités Sci. Indus. 551, Paris, Hermann et Cie.

INDEX

a^{th} coordinate, projection on the ath coordinate, 71
associated sequence of coverings, 51

base of a stack, 13
basis, 27; for a struct, 59; for a uniformity, 56
binary covering, 31
bounds: upper, lower, 4

cardinal number, 9
Cauchy mapping, 62
Cauchy phalanxes, pair (of phalanxes), 63
closed set, 27
cluster point, 34
cofinal, 10; cofinal types, 12
cofinally similar, 11
compact, 37; compact struct, 60
complete struct, 64
complete trellis, 5
completely regular, 58
completion of a struct, 64
continuous, 28; uniformly con tinuous, 61
continuous image, 37
covering, 31; associated sequence of coverings, 51; basic binary, 48; binary, 31; equivalent, 44; finer than, 57; finite, 46; intersection of, 43; large, 59; normal, normal sequence of, 46; refinement, 43; star-finite, star-finite collection of, 46; union of, 43

decided about, 32
decides about, against, for, 32
discrete, 28
disjoint, 2

equivalent coverings, 44

factor, inessential factor, 7
finer than (of coverings), 57; (of topologies), 24
finite, finite character, 7; covering, 46; restriction, 7; set, 3
fully normal, 53

homeomorphic, homeomorphism, 29
hyperslice, 71

inessential factor, 7
infimum, 4
inflated phalanx, 17
intersection, 2
irreflexive, 3
isomorphic, 3
iterated star, 44

large covering, 59
largely compact, 60
linear ordered system, 3
lower bound, 4

mapping (Cauchy), 62
maximal, 7
meets, 2
meshed base of a phalanx, 17
metric, 51
nbd: basis, sub-basis, 24; open, 27
normal covering, normal sequence of coverings, 46
normal space, 29

open nbds, 27
open set, 26
ordered system, irreflexive, isomorphic, linear, properly ordered, reflexive, symmetric, trivially ordered, vacuously ordered, 3

parallelotope, 75
partition, 31

phalanx, 17; Cauchy, Cauchy pair of, 63; base of a, inflated, meshed, 17; of rank n, 65
pointwise product, 72
product, 6; the product, 7; topological or pointwise, 72
projection on the ath coordinate, 71
properly ordered, 3
pseudo-metric, 50

refinement, 43
reflexive, 3
residual, 10
restricting, 45

space, 24; discrete, 28; normal, 29
special pseudo-écart, or spé, 50
stack, base of a stack, 13
star, iterated star, 44; star-finite covering, star-finite collection of coverings, 46; star-refinement, 45
struct, 58; basis for a, 59; compact, largely compact, 60; complete, completion of a, 64
sub-, 2
sub-basis, 27; for a uniformity, 56
subcovering, 31
subphalanx, 35
subspace, 26
substruct, 59
subsystem, 3
superspace, 26
supremum, 4
symmetric, 3
system of subsets, 3

T_1-space, 29
topological product, 72
topology, 24
topophalanx, 33
transitive, 2
trellis, complete trellis, 5
trivially ordered, 3

ultimately, 32
ultraphalanx, 32
uniformity, 55; basis for a, induces a, sub-basis for a, 56
uniformly continuous, 61
unimorphism, 61
union, 2; of coverings, 43
upper bound, 4

vacuously ordered, 3

Zorn, 7

SYMBOLS

$\cong$, 3
$(\Gamma, a^c|c)$, $P\{a^c|c\}$, 7
$\sim$, 11
$>$, 12
$X(\gamma|c)$, 17

Lightning Source UK Ltd.
Milton Keynes UK
UKHW010758120622
404255UK00002B/202

9 780691 095684